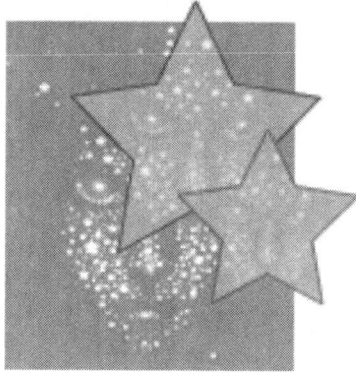

The ORION REGRESSIONS

BY STAN ROMANEK

The ORION REGRESSIONS

BY STAN ROMANEK

DreamSpeaker Creations, LLC
Boulder, Colorado • 2011

All of this story is nonfiction and based on actual events. Some names have been changed to protect individuals' privacy. In some instances, pseudonyms are identified.

First printing: April 2011

Publication Data
        Romanek, Stan
        Hardin, G.W.
        The Orion Regressions
                pp. 220

        ISBN 978-1-893641-06-8
        1. Unidentified flying objects—Sightings and encounters —
        United States—Biography 2. New consciousness 3. Angels

Cover design: GW Hardin

Book design: GW Hardin

Set in 11 point Optima typeface

Dedication

To R. Leo Sprinkle, Ph.D., for his selfless acts of kindness, love, and generosity. Thank you, Leo, for giving me courage to face the hidden truth, as well as the means to embrace the truth when it was revealed. I am deeply indebted to you for helping me accept that I am not a weirdo. Thank you for empowering me with the understanding that the information contained in these regressions is from out of this world. Dr. Sprinkle, you are an asset to humanity!

Acknowledgments

To my good friend and *New York Times* bestselling author, G.W. Hardin, for his unending devotion in helping me put together and publish this book. Thank you, GW, I couldn't have done it without you.

To my wife, Lisa Romanek, for helping me come to terms with these regressions and for supporting me through all this strangeness.

To friend and psychologist, Stanislav O'Jack, Ph.D. Thank you for your guidance and wisdom, especially when I didn't know where sanity could be found.

To Lucie Blanchard, for her friendship and willingness to drop everything at a moment's notice for us, at the cost of many light bulbs.

Heartfelt thanks to all my friends, family, and the researchers involved with my case, for their dedication and support!

Contents

The
ORION
REGRESSIONS

Preface

As a personal eye-witness to and with a bodily-presence in one of the "regression sessions" involving Stan Romanek—a session-process instituted and guided by the notable psychologist Dr. Leo Sprinkle, who has performed a few thousand of such sessions with numerous individuals—what I observed was that Stan's speech changed, in that his vocabulary was somewhat enhanced and the sentence syntax—its structure—was different from that of my many previous conversations with Stan. The timbre of his voice was somewhat altered, and his speech pattern was significantly strange in that his speech was broken up into bits and parts as normally I would, and do, associate that speech peculiarity with someone who is translating a speaker's message to someone else for whom the language is quite different.

The voice deemed or termed "Grandpa" in Stan's paranormal communications is that of a telepathically broadcast message. The said-to-be transmission, which Stan received, is one wherein the essence has to do with humankind's ability to accelerate its progression from adolescence to maturity. That progression is conjoined to a necessary timing so as to call out to our human spirit and, suchwise, to acknowledge that even advanced civilizations are wondering about our paranormal-like spiritual abilities. How wonderful it could be, and will be, when, and only if we would recognize the Mystical Source and Its messages. At that moment, Science (the left-brain dominance) would then unite with the Religious Mind (the right-brain dominance) in a manner where humankind would ultimately achieve a station or orientation of a Religious-Science and/or a Scientific Religious mentality: a "concretized spirituality."

Cordially,
Stanislav Gergre O'Jack, Ph.D.
April 2011

Chapter 1

REGRESSION SESSION ONE

SEPTEMBER 30, 2006

I n *Answers: The World's Most Documented Extraterrestrial Contact Story Continues*, I write in Chapter Three about a series of hypnotic regressions conducted by Dr. Leo Sprinkle with me. Chapter Three is an encapsulated version of all five regressions, with significant editing to make the chapter more readable. However, when it later became evident that the wealth of information proved greater than anyone realized at the time, we decided to produce a complete and unedited version of the five regression sessions. What follows is an excerpt from chapter three of *Answers*:

> ... Lisa and I decided that perhaps we should take a different approach to having me undergo hypnosis again, especially since

several months had gone by since the last session. We explored possibilities of finding someone who specialized in abductees, and came upon the credentials of Ronald Leo Sprinkle, Ph.D., who had received his doctorate in psychology at the University of Missouri. Along with Dr. John E. Mack, a Harvard psychiatrist, Dr. Sprinkle had been one of the first credentialed academic figures to research and study the phenomenon of alien abductions. Founder of the Rocky Mountain Conference on UFO Investigation, Dr. Sprinkle looked like just the person for me to see. And on September 30, 2006, we scheduled the first of many hypnosis sessions with Leo.... As in the previous session with Deborah Lindemann, my personality changed soon into the session. It occurred when Leo tried to find more detail on the advanced physics equations I had written earlier in my sleep. While Leo asked questions about the equations, one of the attendees decided to take a picture because I never opened my eyes during the entire session. Yet I was able to receive pen and paper from Leo as if I could see, and write equations with my head turned aside, never even looking at what I was writing. Leo realized a change had occurred in me and decided to find out why....

Everyone could tell that the speaking voice was not mine, nor was it some aspect of myself. This didn't rattle Leo because he had witnessed other hypnotic regression sessions where another consciousness was able to speak through the client. When I saw this on the videotape, my first reaction was to ask myself, *What in the hell is going on? Who is this? Who is Grandpa?* Everyone in the room initially thought that "Grandpa" was the Grey I had caught on videocam in my house, whom we nicknamed "Grandpa Grey." We gave him this nickname because of his wrinkly skin and wizened look. And it's important to make this clear—"Grandpa" isn't Grandpa Grey. In fact, what we would find out months later is that this consciousness came from an Orion and not a Grey at all.

Dr. Sprinkle began the session by laying the groundwork to see if more information could come forth about the highly technical physics equations I had previously written in my sleep.

Dr. Leo Sprinkle (Leo): First of all, if it's appropriate, ask inwardly whether information about the equations is available at this time. At this time you can speak or, if you wish, you can write. What information at this time is available regarding the equations?

Stan Romanek (SR/Grandpa): Missing piece ... one missing piece ... missing small piece.

Leo: Small piece?

SR: To the last ...

Leo: To the last equation?

SR: Yes.

Leo: Have you that information available to you now?

SR: Yes.

Leo: And can you speak it or write it down?

SR: Either. (*Dr. Sprinkle gives me a pen and paper. I begin to write.*) There's ... symbols ...

Leo: Symbols? (*Reading from my pad as I write*) A2, B2 over N, bracket, line, bracket ... (*long pause*) ... looks like a cube ... energy line. Thank you. Is there any information verbally about the drawing?

SR: No, they won't give me any. It's important ... to the last ... to the last equation.

Leo: The last equation, this is import—

SR: The letters and numbers ... it is the last part of the last equation.

Leo: OK. Part of the last equation?

SR: Yes.

Leo: OK. Is there anything else that you wish to draw at this time in addition to the drawing here?

SR: Not right now.

Leo: OK, now is there information available to you about the source of the equations—the source of the information?

SR: Not sure. They put it in my head.

Leo: Are they willing to provide you with information about the purpose of the contact, the purpose of the equations?

SR: It's to ... I think it's to ... make us work for it ... to make us strive to understand. I don't think they can give it to us ... just give it to us. We have to ... *want* it.

Leo: Can't give us the information directly?

SR: I don't think they're allowed.

Leo: Give us a hint about the ... information?

SR: They ... there are ... laws ... or ... reasonings for not just handing it to us. We have to learn.

Leo: And so it's a gradual process of learning?

SR: Yes.

Leo: Is the information available about the individual equations or the eventual knowledge that will be given through you?

SR: There are others out there that know what this is. It will be given to ... everyone. That's what I think they're looking to do ... it's to get the information out. Pieces will be pieced together. That's how it has to be, yes.

Leo: Other people are receiving information as well as Stan?

SR: That's correct.

Leo: And then that information can be shared with the information Stan receives?

SR: That's correct.

Leo: Is there any information about the mathematics *behind* the equations?

SR: Propulsion.

Leo: That's the purpose, information about propulsion?

SR: It's of propulsion.

Leo: Space-time propulsion ... or ...

SR: Among other things.

Leo: Uh-huh. Are there other types of propulsion that are being given to Stan or others?

SR: Not too much ... a little bit at a time. It's to ... qualify Stan's experiences and to ... help ... lead and guide. Humans are insecure ... humans are still childlike. And ... must be guided.

Leo: And is there a schedule that is being followed in giving information to Stan and others?

SR: Yes.

Leo: And is Stan able to recognize that schedule or be given information about that schedule?

SR: No, he cannot recognize it.

Leo: Is there information right now about that schedule?

SR: It will happen ... soon ... within years.

Leo: Within ten years, five years?

SR: Maybe sooner, it depends.

Leo: Depending on political or social factors?

SR: That's correct.

Leo: And is there any information available through Stan about what those factors might be, political or social?

SR: Yes. Humans are being led down the wrong direction—down the wrong path. And ... humans ... must ... wake up. There is much potential. There is ... a ... enlightenment that needs to happen ... before humans can accept the true reality ... and be accepted into ... *word* ... (*slowly*) *neighborhood.*

Leo: The neighborhood, meaning the galactic or cosmic neighborhood?

SR: That is correct.

Leo: In star nations or star people ...

SR: In your words, yes.

Leo: And [in] what words were they given to Stan—

SR: He is a messenger. He is ... one of ... the few to help ... ease. Humans are scared; humans are led by fear. Humans are ... they are not yet ready. Their understanding is not complete. They are being guided by dark forces ... darkness ... dark forces.

Leo: By dark forces?

SR: That they don't understand ... by human greed and ... that's not their nature ... they are scared. That's why they are being guided by greed. It is fear.

Leo: And the information that's coming through Stan now ... is it appropriate for Stan to speak about the source of this information? Does Stan know?

SR: He has an idea, but he's not sure. He is ... still skeptical—he is ... confused, but he is strong.

Leo: So he'll do all right with the information?

SR: Yes.

17

Leo: And is it appropriate at this time for Stan to be told more about the source of this information?

SR: In time. He must not be overwhelmed.

Leo: Any information today?

SR: Other than what is given? No.

Leo: Is there other information available through Stan about future events?

SR: Yes.

Leo: What information is available?

SR: You are carving ... humans ... humans are carving ... damage. Humans ... are ... creating their own chaos. They are being led to create their own chaos ... for gain of evil or ... *word* ... money ... monetary ... for monetary reasons. And ... humans are still very tribal. And ... wars ... are falsely being ... waged ... lies ... and ... over monetary reasons. And this planet is ... on the brink of ... no return. And we have ... chosen to speed up ... the learning process. So there is a chance to possibly correct the damage and the race ... can save itself.

Leo: And when the race is ready to save itself from this damage, what happens then?

SR: Humans are at a crossroads. They are being judged, and they are being guided ... some bad, some good. There are those on both sides that do not see the benefit of the human race succeeding, and there are those that do ... and that will argue the point.

Leo: So your groups ... you are supportive of humans, and there are groups who are not supportive?

SR: Yes. Just like there are groups of humans who are supportive and there are groups of humans that are not.

Leo: Is there a struggle going on in these outside groups?

SR: Yes, on Earth as well. Stan and his family ... Lisa and the children have encountered this already.

Leo: So the struggle occurs *on*, and the struggle occurs *beyond* Earth?

SR: Yes. But Stan is strong and he will succeed, and he seems to have a good support system here.

Leo: So he can continue as a messenger?

SR: That's correct.

Leo: Is there information available through Stan now about how this struggle is going between extraterrestrial groups or star nations?

SR: It must succeed. For the human race to succeed, it must succeed.

Leo: Is there any information about how that struggle is going?

SR: It is ... undecided on which way, but it must go positively. Humans must learn and learn fast. They are running out of time.

Leo: Is there any information about what those around Stan can do to be helpful?

SR: When the time is right it must ... the information must get out. There are others waiting. At all costs the information must get out. There is going to be opposition, but there are a lot of people ... beings ... behind Stan that support him and watch after him... even within your own government. This message is for everybody, though— for covert groups, for people of evil intentions, for the people who spy on Stan and his family—it's for them also. It's for everybody.

Leo: Is Stan going to be given information about contact with the source of this information? Will he be told eventually the source of this information he receives?

SR: He knows. Inside he knows ... where it's from. It is still hard for him to accept. And ... he will get over that.

Leo: Is information available about whether the source of information not only is extraterrestrial but also extra-*dimensional*? Are you coming from—

SR: It is the same.

Leo: The same? There are some individuals who are flesh and blood like humans?

SR: That's correct.

Leo: And there are some who are intelligences without [human] minds?

SR: Yes.

Leo: And are these groups working together?

SR: Some are. Some are beyond the need for that. Some are such that they don't worry about things.

Leo: Or worry about the struggle here on the planet?

SR: No.

Leo: Is that the main source of the information about propulsion systems—does that come from extraterrestrial groups?

SR: It comes from what you call "ET."

Leo: ETs?

SR: ETs, yes.

Leo: Do you describe yourself as ET, or do you describe yourself as "star person"?

SR: You describe us as ET. We are from somewhere else ... far away.

Leo: From another dimension?

SR: No.

Leo: From another planet?

SR: Yes.

Leo: Is information available about that planet?

SR: Stan has already received information.

Leo: OK ... and is there further information right now that would be helpful to provide for Stan.

SR: It has ... been given to the right people; they have figured it out. It has been given to three others.

Leo: Like Stan?

SR: [There is] one other that they know of.

Leo: Three other persons?

SR: That's correct.

Leo: And does Stan know who those people are?

SR: No.

Leo: But the information has been given so that there is more than one messenger available?

SR: That is correct; one of which they know of ... meaning, not Stan, but whom Stan is working with.

Leo: Is the name of that person available now?

SR: Yes.

Leo: And who is that person?

SR: (*Slowly*) Susan ... Susan.

Leo: Susan.

SR: Susan.

Leo: Is Susan in this area?

SR: No, she is on the eastern part of the continent. (*Slowly*) Carson ... Carson [pseudonym].

Leo: Carson. OK ... anybody else that is receiving this information?

20

SR: She knows of whom they talk about. She knows the researcher that knows this, one of the people specifically.

Leo: So that comparisons can be made between information received?

SR: It has already been done.

Leo: Already done. Is information available today through Stan about other topics, [such as] crop circles or other sources of information?

SR: Yes.

Leo: What information is available right now?

SR: It is a … learning process. It is a way to evolve the thinking of humans. Humans think linearly. They cannot think outside of their line. They cannot think off to the left or off to the right; they must be taught how to do that. It seems to us that they have been guided incorrectly, but we do not know who, or how, or what. There is a … a … influence not of their own doing. Humans are … not altogether from this location; they are a conglomerate … conglomerate of many, and they have been watched for a long time. But it is time for them to grow up now.

Leo: Is information available through Stan now about human origins? Have there been many planetary civilizations here that have settled on planet Earth?

SR: There have been visitors, and there will continue to be visitors. But it is for the good of man not to rush into it. They are not ready—they are not … *word* … social … socially accepta … acceptable of what is really out beyond their sight … and there is a … ongoing … there is an ongoing … *word* … um … (*exhaling*) … design by … the ones in charge now to keep them … misinformed and keep them quieted … for monetary reasons.

Leo: Talking about humans … you're talking about—

SR: Yes, humans …

Leo: Other groups beyond the planet?

SR: Hmm … *some* groups beyond, but mostly [those] who [are] ready. Governments know. It is not what people think. It is a … *agreement* between governments … so the human race can be kept quiet … and for monetary and for selfish reasons.

Leo: Is there information through Stan now about what kind of procedures will release that information within the next few years?

21

SR: It will get out to Stan the way it has been. It cannot be "given." It is just ... it is used to guide ... to guide. And ... there are those on this planet that know already, then refuse to share it. So it [has] been given a different direction for all. Information is already out there ... but those who want to control, control by lies ... and refuse to enlighten—because of monetary reasons.

Leo: Thank you. Is there information about the possible schedule? Does Stan know about the possible schedule on releasing information?

SR: No. It is determined on ... it depends on what happens from now on.

Leo: What happens on the planet?

SR: Yes.

Leo: [With] world governments and world leaders?

SR: Yes.

Leo: Is there information about further equations that Stan may be given? Or if you could, give him further information about propulsion systems.

SR: If needed. As it is needed, or determined that it is needed.

Leo: Is there more than one propulsion system for spacecraft?

SR: Yes. Yes, many. There is already information here that ... they will not share it, so the information has to get out for everybody. It is to help ... Stan's experience is to help ease [the] human race into acceptance of what *is* ... out beyond their reality. And they must learn quickly. They are running out of time.

Leo: Can Stan's experiences help other people recognize that there is assistance, that there is support from other groups?

SR: Yes ... in a subtle way, yes. But it is more. Stan is a ... he is a ... *word* ...

Leo: Example?

SR: Yes, he is a ... example. And ... he is also a ... *word* ... a ...

Leo: Role model, or a—

SR: Yes, exactly ... role model, but he helps those who guide him also. This is not just for humans; this is for ... everyone. This is for ... the ones with evil intentions. This is for ... everyone. Stan knows this; Stan talks about this all the time. He knows it in his head because it is true. Everyone must learn; everyone must know.

Leo: Everybody grows up.

SR: It is time to grow up. The human race is at a crossroads. They must know that if they continue the way they are, they will not be here long.

Leo: Is there a contingency plan if humans don't survive the planet. Does Stan know any information about a contingency plan?

SR: No. It has not been decided. A race must learn on its own. A race must bring itself up. Humans want to have the right to ... sovereignty; they want to have a right to live the way they want to live—they must prove themselves. They are a ... immature, insecure race right now. They are ... growing, but they have come to a crossroads where if they proceed out into the immense, they are not ready. They could cause harm to others, and that will not be allowed ... until they learn.

Leo: So that violence or war or control of others is not permitted?

SR: It will be ... maintained so that it does not reach past this planet. They must learn—because if they do not, they will destroy themselves ... and everybody knows that.

Leo: And are you willing to provide Stan with more information about your own knowledge, your own intelligence? Do you speak as one entity, or do you speak as many?

SR: There are many. We are just a small group. We monitor and watch and ... if possible, provide assistance. Stan is special. (*Turning head toward Lisa with eyes shut*) Lisa is special. The people that are involved with this, including yourself, are here for a reason ... you have been chosen. All of you have been chosen.

Leo: Are you willing to tell us more about your home planet—your civilization?

SR: Some.

Leo: What information are you willing to share with us?

SR: It is beyond your understanding in most parts. We do not bicker; we do not have war. Primitive cultures fight with themselves. Humans are ... special. Humans seem to have the ability to relate to other creatures and can coexist. Although there are other instances of that, humans are best at it and it is rare. This ability is ... this ability is ... so strong in humans.

23

Leo: Is there other information about your civilization [inaudible]?

SR: A little.

Leo: Have you monitored humans for a long time?

SR: Yes.

Leo: Do you have information about human origins, about how humanity began on Earth?

SR: It was a ... not by us.

Leo: More advanced?

SR: As advanced as we are from you, they are as advanced from us. Humans are a conglomerate; they are part of the planet you call Earth, and there are more. There are offshoots that do not belong with humans but in the struggle to ... to find the origins, humans have ... accepted that they are from ... *word* ... from this particular species [apes]. In fact, they are a conglomeration of off-world and on-world ... older than they think. They existed when what you call ... cave man ... existed with cave man ... in smaller groups—in little pockets.

Leo: In other planetary systems?

SR: They were designed to inhabit and survive here. And ... we did not understand; we did not ... have any part in this. A much more sophisticated intelligence did. And ... they are ... we do not know why.

Leo: Is there a major group of leaders off plant Earth who are sharing information with your group? Can we follow suggestions or advice from other—

SR: It is a conflict. There are those that do not care or do not think it is wise, but we believe that humans are more than [these others] understand and ... humans seem to be more than *we* understand.

Leo: So you are supportive and willing to help humans, as you wish to see them grow up or advance?

SR: Yes. That is why I am here.

Leo: Do you have a name or a title?

SR: It is not important.

Leo: Others around you—do they have a name or a title?

SR: They do but it is not important. They have contacted ... our ability to use human technology is inefficient because we can only do it at

24

certain times and because of location. We have foreseen certain things and have contacted Stan and [other] people involved regarding theses things. Stan is not alone. He is being assisted, and he has agreed to help. It is hard for Stan to accept what he has encountered and is encountering, but he is special and does not yet understand ... in that way.

Leo: And will more and more information be given to Stan—that he can work as a messenger?

SR: As it is needed. We do not mean to overload or to frighten. We will give things as they can be accepted and withstood. We will make visits as is needed. Contact has already been made. And it will escalate because we are running out of time.

Leo: Is it all right for questions from others to be asked of Stan?

SR: Yes ... I must go. I've run out of time. I must go.

Leo: We'll continue when you come back later

At this time I awaken; Dr. Sprinkle tells me it's time to take a break. Then the session resumes.

Leo: OK, Stan, you can be open to information that comes from a benign source, allowing the information to come through you—information that's meaningful and significant for you personally, but also significant to family and friends, colleagues, those with whom you are working—knowing that for whatever reason messages are being given to you, that you can share this information with those with whom you're working, etcetera. And so we seek information through Stan from these sources of knowledge ... etcetera. Is it appropriate to ask questions and ask Stan to speak?

SR: Yes.

Leo: Is it appropriate to ask why Stan is "special"? Is that information available through Stan?

SR: I don't know.

Leo: OK. Is information available about why Lisa is special?

SR: I don't know.

Leo: Is information available now about Stan's leg or information about the condition of Stan's leg?

SR: It's fixed.

Leo: And is there information about how it was fixed?

SR: They stuck needles ... not needles, but ... [it] almost looks like light ... rods of light.

Leo: Rods of light?

SR: M-hmm.

Leo: And rods of light provided some kind of energy that was healing the leg?

SR: It hurt. They stuck it in, but, yeah, they fixed it. Lots of pain—but they made the pain go away.

Leo: Is there any other information about the condition of the leg in the future?

SR: It was hurt pretty bad, but ... they fixed it.

Leo: Is there information about Stan's blood sugar level? This condition—medical or biological condition?

SR: I don't know.

Leo: Is there anyone we can call upon to ask for information about [Stan's] blood sugar level? Can somebody speak through Stan?

SR: Yes.

Leo: Is there information about blood sugar?

SR: It is ... it is genetic. It is genetic.

Leo: It is genetic? And is there information about what can be done to ease the condition or to assist Stan to be healthy?

SR: It is a combination and ... he will survive.

Leo: He will survive—does he know what he must do to be as healthy as possible?

SR: No.

Leo: Is there information that would be helpful to Stan about diet, exercise, or medications? What would be helpful?

SR: Stress.

Leo: If he minimized his stress—that would be helpful?

SR: Yes.

Leo: And are there suggestions about how Stan can be helped to deal with stress?

SR: No. He must go through this ... as part of the process.

Leo: Go through this work that he's doing?

SR: Yes. It's stressful.

Leo: It's stressful but he can learn how to minimize stress and still do the work?

SR: Yes. It is not all from stress. It is … part of … it is genetic and caused by injury … and illness. It is a combination.

Leo: Injury and illness?

SR: Yes. It will dissipate, but it will not go away permanently. It is bad because of overwhelming stress. But he has a good support system.

Leo: And so if he is able to run the support system, if he is able to maintain his health, he can continue to do his work?

SR: Yes. We will make sure of that.

Leo: Did you help with Stan's leg?

SR: Yes.

Leo: And what did you do to help Stan's leg?

SR: Fixed.

Leo: Did you use a special technique or special procedure?

SR: Not for us.

Leo: Someone else?

SR: We used what is used for us.

Leo: Oh. And can you describe that process?

SR: It is not important.

Leo: The fact that it's fixed—*that's* important?

SR: Yes. He must proceed, he must … not waste time … it's too important. He must be well able to … proceed.

Leo: Is information available to Stan about why he is important? What happened in the past to make him special or make him important?

SR: He was chosen. He is … right. He is … smart in a different way. He is strong in a different way.

Leo: So his smartness and strength is useful to continue to work as a messenger?

SR: And … he has been enhanced.

Leo: He has been enhanced?

SR: Yes.

Leo: By changes in the body or in the mind?

SR: Yes.

Leo: And is it appropriate to ask what changes have been made?

SR: He … can understand, where others cannot.

Leo: He can understand the information that's being given to him?

SR: He can process what is happening more efficiently than others. He can guide. He will lead with efficiency. Everyone has their part; this is Stan's part.

Leo: Stan's part ... as messenger—he's getting information about propulsion systems? That's one part?

SR: That's a small part.

Leo: Is there another part—

SR: Yes. He will ... speak about his experiences so that others may learn. And he will help ... with the transition.

Leo: That was "transition," for others?

SR: Yes. And you will also help with the transition.

Leo: Others will help ... you will help as well as others?

SR: Yes.

Leo: And is there information about the schedule of this transition?

SR: It depends on what happens. It depends on the direction.

Leo: OK. Is information available about Stan's family?

SR: Yes.

Leo: Are they involved?

SR: Yes.

Leo: What is their role?

SR: They are of interest because they are with Stan ... and that is all.

Leo: Is there information through Stan right now that will be helpful to Lisa and [Stan's stepchildren]?

SR: They are ... twins. Girls are twins. (*Long pause*) Fascinating. They are smart.

Leo: They are smart?

SR: No connection ... just ... they are with Stan.

Leo: No connection with a specific task?

SR: No. They are with Stan,

Leo: Are there special tasks for Lisa?

SR: She is supporting Stan. And she must ... we ask ... that she finish ... because it will help ... in the goal.

Leo: And others, Robert, Alejandro, colleagues of Stan?

SR: Yes, they all part ... they are supportive and they are part ... not by accident, they have been chosen also.

28

Leo: Is there special information for them right now that would be useful for them to know?

SR: Just to continue ... what they do. (*As Lisa begins to ask Dr. Sprinkle a question, I turn my head toward her with my eyes still closed and speak to her.*) You can ask me.

Leo: Do you have a question, Lisa, that you want to ask?

Lisa Romanek (LR): How would your sister, brother, and your father—

SR: They are in the same genetic line. They are ... part of the process to get to Stan. They are ... they have been tested; they have been ... it has been determined ... who is correct, who is right, who is the one ... to proceed. It is Stan. It is ... his job to ... among other things, it is his job to help with the transition.

Leo: Is the transition also part of the communication from ... others, those from [unintelligible], [Audrey]? Is there information about these sources of information?

SR: Yes. They are ... efficient in human technology, but we cannot ... communicate unless things are right ... unless things are ... because of distance, because of ... there are factors. We can communicate only when the window of opportunity is available.

Leo: And that's why voices over the telephone are used to communicate?

SR: It is easier for them to accept. Although some do not believe in the messages, it is easier to accept than other forms of communication.

Leo: Will these forms of communication continue toward the [unintelligible] the transition?

SR: If needed. If it is important, then it will be communicated.

Leo: Is there other information available about the transition?

SR: It is ... only acknowledgment that there is more ... beyond their ... beyond their ...

Leo: Understanding or their knowledge?

SR: Uh ... beyond their ... view.

Leo: Beyond their perspective?

SR: There is more ... more ... many more ...

Leo: So if Stan's family and Stan's friends, colleagues, accept that there is more ... this would be helpful?

SR: Yes, if not just Stan's family ... if everyone on this planet, everyone ... humans must accept ... they are ... Stan's words ... narciss ...

Leo: Narcissistic?

SR: And they think about themselves and do not understand there is so much more out there. They must, in order to be accepted, in order to save themselves, they must accept the fact that they are not ... they are not by themselves.

Leo: So if humanity, in general, accepts that we are not alone ... that's good?

SR: It must be ... yes, it is good. It must be ... it must be accepted everywhere. It—

Leo: Everywhere on the planet?

SR: Yes. It ... even though when the shift happens, we will know— when everyone will accept. We have tried suddenly to introduce more and more and more to ... get this race evidence. It is for their own good.

Leo: So awareness of civilizations outside Earth, that's important?

SR: *Accepting!*

Leo: And in accepting, you mean acknowledge?

SR: *Accepting* will bring knowledge, accepting will bring awareness, and that is our goal. Our goal is to bring ... acceptance.

Leo: So awareness, acceptance, and acknowledgment, publicly, are the goals?

SR: Yes. Then you can help. We cannot help other lives, we cannot ... trust who's in charge. We do not trust ... humans at this time. They are too ... immature; they are too tribal. They are like—because of fear. They must grow up ... for their own good ... and it must happen soon.

Leo: Is information available about orbs and flashes or light that sometimes are around Stan?

SR: Orbs?

Leo: Uh ... balls of light? Flashes of light?

SR: On ... flashes of light is ... (*slowly*) ... *ways ... word* ... their ways ... their ways ... we monitor.

Leo: Monitor Stan, monitor—

SR: We have mistakenly ... fear. We have mistakenly ... been ... *word* ... lazy, *not* lazy ... not attentive to our task and have been caught ... by Stan and Lisa.

30

Leo: And his friends are aware of the orbs, they are witness to—

SR: Not orbs, we have been ...

Leo: Photographing? Photograph, videotaping?

SR: Yes. Some we allow for ... confirmation, mostly allow for confirmation.

Leo: Oh, so that Stan's friends and other people can see that what Stan sees are real.

SR: That's right. Sometimes we ... hah ... are caught off guard.

Leo: Oh. So sometimes it's intended and sometimes not! (*Laughter*)

SR: Yes. (*Laughter*)

Leo: Is it happening now?

SR: Is it intentional? Yes.

Leo: Have orbs, lights now ... if photographs being taken now of these lights be [unintelligible] ...

SR: Orbs. Orbs ... orbs are not what you think. There are red orbs that are surveillance. There are ... some that are just ... orbs. There are some that are just ... there are some that are part of us, and there are some that are just part of the ... (*slowly*) environment.

Leo: Part of the environment?

SR: Yes.

Leo: And so some of them are intentional, directed by you, and some are part of the environment?

SR: Yes.

Leo: Thank you. As you get information about Stan and his work, do you also give information through Stan about your background, about who you are?

SR: Some. Some have ... information. Some already have information.

Leo: Are you able to talk through Stan about your civilization, about your language, about your customs?

SR: Hmm ... a little ... a little bit.

Leo: Can you speak through their English?

SR: We do not speak like you do. We do not ... you are primitive still. You are primitive ... you speak ...

Leo: Speak with a voice?

SR: Yes. We do not need to.

Leo: You speak mentally?

SR: Most advanced or more advanced civilizations can do this in long distances. It is easier to ... it is easier to ... (*slowly*) *explain* ... it is easier to portray ... what needs to be done next.

Leo: Mind communication rather than vocal communication?

SR: Yes.

Leo: Are you able to let Stan know mentally, telepathically ... information?

SR: Yes, that's how he gets the information.

Leo: With his conscious awareness?

SR: Sometimes. If needed. It is a matter of ... overwhelming him or not. Humans also have this ability that ... they do not understand ... it has been ...

Leo: Damaged?

SR: No. Forcefully ...

Leo: Suppressed?

SR: Suppressed ... *word* ... suppressed.

Leo: Forcefully suppressed so that communication is available and can be hoaxed and lied?

SR: We do not understand. We are just ... we don't understand. We are ...

Leo: So that if humans could enhance our type of communication, that would be positive? That would be good?

SR: It is automatic. We believe it would automatic ... with their awareness ... of ...

Leo: Knowledge? Truth?

SR: No ... of ... we believe it will be a trigger ... a trigger, (*slowly*) *trigger* ... (*slowly*) *when* they are aware.

Leo: Is technical communication available to enhance telepathic communication?

SR: It should be natural. It *should* be ... it is slowly being regained. There are many like Stan and like his sister, and even Lisa, a little bit ... and even *you*.

Leo: Thank you.

SR: And ... that it will slowly get better, but when there is an acceptance of ... when the human race accepts that they are ... not alone, it will, we believe, it will be triggered.

Leo: Enhanced?

SR: Yes. We believe they will be triggered, too.

Leo: Is there anything that you wish to tell us that might be helpful to Stan so that he is able to accept his work and not be stressed by his work?

SR: He is not stressed by his work. He is stressed by what other people think of him and ... he is stressed ... he is ... intelligent in this way— he can understand what his goals are, but he does not really understand himself and who he really is. He ... ha-ha ... ask meaning ... he ... is ... different.

Leo: Accepting his difference—is that part of the task?

SR: Yes. That's part of his task. Humans are here for a reason. They must be here. And ... that's all I can ...

Leo: So if Stan not only accepts his work, his task, but also accepts his own part in it, this will be helpful, regardless of what other people think?

SR: He has accepted his part, but he does not understand who he really is ... or *what* he really is.

Leo: And in terms of his past ... or his ...

SR: His past and future and present.

Leo: Is information available now that might be helpful to Stan when he is able to—

SR: It will be given to Stan.

Leo: Is there other information that you can share now about your work with Stan, how long you've been with Stan?

SR: He is ... part of us. We are part of him. He is a guide; he is a messenger. He is here to help. Everyone has a purpose on this planet. This planet is unique in the fact that you have places ... this planet must grow up, though. Must grow up. Everyone around Stan has been chosen for a reason. Everyone in the past that is no longer with Stan has been chosen for a reason, as small as it is ... to Lisa, who is most important in supporting Stan.

Leo: Thank you. Do you have information about what humanity calls the Greys? Are they good for humanity, the Greys?

SR: The Greys are ... we are ... they are ... on Earth as ... there is ... there are some that want ... this to succeed, and some that do not.

Leo: So there are different groups.

SR: That is correct.

33

Leo: Do they come from a specific planet or a specific civilization?

SR: They are in many places now, yes. We are in many places.

Leo: Are you from the Zeta Reticuli?

SR: There are some from there, but we are not from there.

Leo: Can you tell us where you're from?

SR: (*Slowly*) *Susan* Carson ... Stan knows ... Susan Carson knows. We are past ... *word* ... we are past ...

Leo: Past a particular part of the star system?

SR: Yes. We are ... it is in the way. When you look for ... you cannot see us. Something else is in the way.

Leo: The astronomers cannot even see?

SR: They can, but something else is in the way. Um ... it is ... two stars ... three stars ... following to, not seen very well from here.

Leo: Is it the Orion Belt?

SR: It is Orion.

Leo: Orion?

SR: Yes.

Leo: Thank you. Is Stan helping you as a race? Is that part of his work; is he helping you create children?

SR: Yes, he is helping us understand.

Leo: Understand, or is he helping you create children?

SR: Yes, to understand.

Leo: As hybrids?

SR: Hybrids? Yes.

Leo: Is that how you help humans on the planet?

SR: It's to help everyone.

Leo: Not only here on planet Earth but elsewhere?

SR: Yes.

Leo: Is there other work besides hybrids that Stan is helping you with?

SR: He is ... helping here. He is ... this is for everyone. This is for ... you; this is for them. This is for ... everyone. His experiences will help with the transition. His experiences will ... among other things, will ... ease that fear. That fear ... ease that fear.

Leo: Ease fear of many races, many cultures, [unintelligible]?

SR: They are fighting among themselves. They are ... too many. They are being led incorrectly. They are ... they must take the focus off themselves.

Leo: So the transition is not only to help our planet Earth and humanity, it is also for other civilizations?

SR: Yes. If they can ... and Stan knows this ... if they can get the focus off themselves, it will help in the healing. There are forces who are ... um ... they are led by ... monetary ... they are selfish ... these forces are leading them in the wrong direction ... for their own personal gain. And they will be dealt with.

Leo: Did Stan and his family have a special role? Did Stan's father have a special job?

SR: He was ... involved without knowing it. He was ... the first step. His family was monitored ... and Stan is monitored ... because ... they are part of Stan.

Leo: Monitored to see who might be able to serve and work well [unintelligible]?

SR: Hmm ... no, Stan is different.

Leo: But the family is still being monitored?

SR: Yes. So is Lisa—and the children because they are with Stan.

Leo: We appreciate your comments, and you seem to have a sense of humor about being photographed. Have you been photographed before?

SR: Mistakenly, yes, (*whispering*) by the fence. (*Slowly*) *By the fence.*

Leo: By the fence. Oh, yes! (*Laughter*) Yes. Thank you. Do you have a name, or is it OK—

SR: It is not important.

Leo: Is it OK if Stan calls you by a name?

SR: If they want to, they can call me "Grandpa."

Leo: They can call you Grandpa?

SR/Grandpa: Yes.

Leo: And then that—

Grandpa: And there's ... there are others like me that are involved also.

Leo: So, you look like you did when you were by the fence?

Grandpa: Yes.

Leo: And that's why it's OK to call you Grandpa?

Grandpa: Yes.

Leo: OK. Is there other information you can—

Grandpa : She will be well in two days (referring to Lisa who was suffering from a cold).

Leo: (Laughter) Thank you. So she'll be well in two days!

LR: (Giggles) I hope so.

Grandpa: Yes.

Leo: She has what we call a "cold."

Grandpa: Yes.

Leo: Is there other information you can share with us about your work and background?

Grandpa: A little.

Leo: What is your main job, your main function?

Grandpa: Observe ... and to guide

Leo: Observe and to guide. Do you have others like Stan whom you observe and you guide?

Grandpa: Some. Stan is ...

Leo: But Stan is the main man?

Grandpa: Yes.

Leo: Stan's the man, huh?

Grandpa: Man.

Leo: Sole person? Is there spiritual information about Stan?

Grandpa: It is all part, it is all part of this, it is all ... it is an awakening; it is a ... gift. If focus can be taken off themselves, they will see how amazing it truly is.

Leo: So that's why you don't focus on your own—you're just part of that work. It's OK for me to call you Grandpa, but do we have to know anything else about you?

Grandpa: I am not important. What I do is important.

Leo: Thank you. Have you been with Stan and Lisa before? Have you been to Robert's cabin?

Grandpa: Others have. I have also.

Leo: You've been—

Grandpa: There are others.

Leo: And that means they work with you?

Grandpa: Yes. Other races.

Leo: Other races?

Grandpa: Yes.

Leo: Do you look different than those groups?

Grandpa: Yes.

Leo: Can you give us information about the other ... races?

Grandpa: (*Slowly*) *Stan calls them the "Possum People."*

Leo: The Possum People?

Grandpa: Yes. They are like Stan and like us, together.

Leo: And if Stan is able to minimize his stress and to be healthy, will his work continue?

Grandpa: Yes.

Leo: Do you know what would be considered success or achievement in his work?

Grandpa: Information must get out.

Leo: That's the main goal, is for people to—

Grandpa: And after that, it will be natural for Stan to help and guide others, how he is.

Leo: Is there any information you can share with us about Stan's work?

Grandpa: He is to be as he is supposed to ... everyone here is chosen to do what they are supposed to do.

Leo: Is information available about these other groups, the Possum People, these hybrids?

Grandpa: They are like us and like you.

Leo: Humanlike?

Grandpa: Yes. They are like us and like you together.

Leo: How about the Praying Mantis, the tall group, are they part of your work?

Grandpa: It is not important.

Leo: Any other groups?

Grandpa: There are many. There are many groups.

Leo: Many groups.

Grandpa: Yes. There are many, many, many ... some acting together, some not.

Leo: Do the different groups have different functions?

Grandpa: Yes. There are up there as there is here. There are some that are concerned with their own intentions, and some that are more spiritual. There are some that are concerned with humans, and some that are not.

Leo: Should Stan be afraid of any of these groups?

Grandpa: Yes.

Leo: Should he be careful about the ones that are selfish or self—

Grandpa: That is correct. Mostly his own kind.

Leo: Mostly the other humans.

Grandpa: Yes. There are some that are not positive ... do not care to see this ... believe this is a waste of time. They believe it is not ...

Leo: So in general, when Stan is meditating or he is asleep, he can be contacted by you?

Grandpa: Yes.

Leo: Is he likely to be contacted by anyone from the negative group?

Grandpa: That could be. We try to avoid that.

Leo: Is there anything Stan can do in meditation to prevent encounters by negative groups?

Grandpa: We have tried to ... guard him. We have tried to ... intervene ... and ... we will continue to do so.

Leo: So if he knows he's being protected and guided, that should help ease his fears?

Grandpa: Yes. We can sometimes control humans, but most of the time we cannot ... as Stan knows.

Leo: Does Stan know that you're guiding and you're helping?

Grandpa: No.

Leo: He doesn't?

Grandpa: He does not know. No.

Leo: But if he feels that [unintelligible] help with his work?

Grandpa: It's not important.

Leo: He'll continue to work.

Grandpa: No matter what, he is driven that way.

Leo: So the main thing for Stan is just being careful of him[self] and his work.

Grandpa: Yes. There are some that are not who they pretend to be. But this is for everyone. This is for those pretend—who, who try to mislead. This is for everyone.

Leo: Are you and your group also protecting Robert and ... Alejandro and—

Grandpa: We try, yes.

Leo: Other members of the family group?

Grandpa: Yes. Anyone that is involved. This must get out.

Leo: Can you explain what you mean when you say that Stan is like you? Is he one of you?

Grandpa: He has part of us ... but—

Leo: Genetically?

Grandpa: Yes. A little bit.

Leo: A little bit. And is also, the work [unintelligible] similar?

Grandpa: Yes. All of the above.

Leo: So genetically there's some connection?

Grandpa: Yes. He is ... his illness ... we *think*, we're not sure ... his illness might be because of his connection with us. But we do know that it's genetic.

Leo: And is there not much that can be done about that?

Grandpa: Not right now. Not right now. He is where he is because of his experiences and his ... he is the way he is because of his goals, and there are those here that do not want that. He is constantly being ... *word* ... constantly being ... (*slowly*) bothered ... um ... *word* ... (*slowly*) har ... harassed.

Leo: Harassed? Do you know if people close to Stan are harassing him or not to be trusted?

Grandpa: Yes, there are one, or two ... that he should consider ... not being around.

Leo: Do you know who would cause trouble, to Stan or his children?

Grandpa: I think they know.

Leo: Is [name deleted] a name, that ...

Grandpa: [Name withheld] is ... [name withheld] is ... unconsciously causing stress.

Leo: And others?

Grandpa: He'll figure it out on his own.

Leo: The main thing is that this is his own work.

Grandpa: This is for everybody. You cannot exclude those that are out to ...

Leo: Is there anything else that you'd like Stan to know before he returns to a normal state that would be helpful to him in his work?

Grandpa: He already knows.

Leo: He already knows. We thank you so much for the information you have shared with us. Thank you very much.

End of regression session one.

Chapter 2

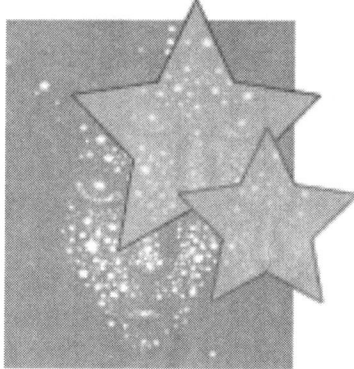

REGRESSION SESSION TWO

NOVEMBER 2, 2006

I n October of 2004 those in our group began getting phone calls and phone messages from a computerized voice we eventually called "Audrey." We called the voice Audrey because it matched the same voice used by the AT&T Natural Voice speech engine found on various speech-driven technologies. Over time we found out that there were three sources using the Audrey voice either to deceive us, help us, or guide us—the Black Ops, White Hats, and ETs. Black Ops used a version of Audrey that did not match the technology used by the ETs. White Hats (Black Ops who were good guys trying to help us) used Audrey to warn us of danger from their fellow Black Ops. This version also did not match the technology of the ETs' Audrey. But we knew when we were being helped by the White Hats. The Black Ops usually used the Audrey voice to deceive us or threaten us.

The ET-based Audrey placed calls with several individuals on October 28, 2006, including two messages to Dr. Leo Sprinkle. The result of those messages was the scheduling of a second hypnotic regression session with invitations to seven individuals who the ETs had identified as people of influence who could contribute their talents in helping our world. Those of my inner circle also attended the session. What follows is the Audrey message to Dr. Sprinkle, played to the attendees before the session began:

> Audrey: Hello, Leo Sprinkle. We apologize; our timing appears to be a bit skewed. We would have preferred talking to you in person. Unfortunately, you are not there, and our ability to contact and visit your world are limited by various factors. We want to thank you for working with Stan. Your help is greatly appreciated. Everyone involved in Stan's case is involved for a reason, and you are no exception. Events have been carefully orchestrated to help Stan's work progress. This is very important and must succeed for humankind. Unfortunately, there are limits to what help is given. Mankind must learn on their own with lowly guise. Your involvement can help Stan tap into a connection that would have otherwise been ignored. Recently it has been granted that more communication is available and a dialogue can be made so questions can be asked. Stan and others will know whom to invite. You will decide how you want the questions to be asked. We suggest that questions be written and given you to be asked. This must happen soon or the window of opportunity will pass. Please let others in the group know.

The session began.

Leo: Welcome. Thank you for being here.
SR: Hello.

Leo: And do we call you "Grandpa" or ... do we have another name for you?

Grandpa: You can call me Grandpa if you wish.

Leo: Thank you. We have a group of people here, who support Stan ... and we have more questions. Some of them may seem repetitious or [unintelligible], and we hope you are willing to respond to these questions.

Grandpa: (*Nodding*)

Leo: It would seem that it was important that we speak again. Is there new information that you have to share with us since our last meeting?

Grandpa: No new information.

Leo: Is there something that has happened or changed to make this meeting important?

Grandpa: You have more questions.

Leo: Yes, we do. Are there any new equations or additions to the equations that you have shared in the past?

Grandpa: No more.

Leo: No more? Will there be future gatherings that will be considered?

Grandpa: It depends.

Leo: Thank you. Now I have a series of questions from Susan Carson [pseudonym], who works with a group of scientists, and I'll ask these questions and invite you to respond if you wish to do so. The first question: Are you aware that the humans who are trying to support Stan are doing so in the hope that your involvement with him is for the benefit of humankind?

Grandpa: Yes.

Leo: Are you aware of the human concept of good and evil?

Grandpa: Of course.

Leo: And is your interest in Stan for his good ... for the good of some people, or for the good of all humans?

Grandpa: For everyone.

Leo: Is your involvement with Stan what you would call "good," specifically for Stan, as well as for humankind in general?

Grandpa: For everyone.

Leo: For everyone. Thank you. We and Stan need to be certain of your motives. What can you tell us about yourself or your agenda that would help us understand why you're involved with Stan?

Grandpa: For the good of man.

Leo: For the good of mankind? OK. And what benefit do you or your group hope to obtain from your involvement with Stan?

Grandpa: Learning. Humans are ... humans have potential. Humans ... must be guided, though. Humans are ... lost ... not of their own doing. Humans are ... being guided incorrectly. We are ... guiding ... so they can be enlightened. We are guiding so they can ... connect with others that are out ... beyond their imagination.

Leo: Thank you. Is there a factor or factors, which make this need for learning more urgent this time?

Grandpa: Yes, there is. It is ... *word* ... self-destruction. It is ... greed. It is ... what you call "evil" that is ... pushing ... humankind to the brink of nonexistence.

Leo: OK, thank you. Now here are a few questions from Lisa and Stan. Stan is interested in the source, your source—who do you represent—and why you call [Stan], "Starseed."

Grandpa: We are ... ones that care ... what happens to mankind. Starseed is a ... *word* ... (*slowly*) a ... *nickname* ... nickname for Stan. He is different. He is a messenger. He is ... he is ... *word* ... he wants to help. He is different; therefore we call him "Starseed."

Leo: Thank you. And Lisa asks why do you or those working with you erase memory from people's minds?"

Grandpa: Why do humans ... tag animals? Why do humans ... study animals? We are no different. Humans are primitive, but ... we know that they have potential. They can create what they imagine. They are ... learning very quickly. We want to ... guide them in the right direction.

Leo: OK, and is that the same response for why you or your representatives are sneaky, peeking between the windows, sneaking around houses? Taking information from people?

Grandpa: Not all ... at this time. Some do to ... study reactions to ... learn ... most that have been taken ... not all are positive. Most are positive. Most, we try to ... eliminate memory because ... the experience might ... not be pleasant. But it is necessary for us to learn. It is necessary for ... *word* ... (*slowly*) exp*eriment* ... experiment.

Leo: So learning about humans and experimenting with humans is one of the motives?

Grandpa: Yes.

Leo: Yes. Here are some questions from Alejandro Rojas and his colleagues: Do all people that you contact—abductees, contactees, experiencers—do they give consent for these experiments?

Grandpa: (*Pause*) In ways you do not understand. Stan's ... connection with us is genetic and ... he has agreed subconsciously. He knows it's for the betterment of mankind and ... to help us.

Leo: And the entities known as Greys, do they ask for consent?

Grandpa: Some do; some do not. Some do not have the same motives we do. We are not ... of them. There are [some] among us that look alike but do not have the same motives. Some humans are taken against their will ... and some agree.

Leo: In what ways do people give their consent?

Grandpa: In ways you do not understand—unconsciously and ... *word* ... generational.

Leo: Generational?

Grandpa: Generations. Generations.

Leo: Does that include previous lives?

Grandpa: Family units, mostly.

Leo: And so that's why some people don't remember ... giving their consent?

Grandpa: Yes. Some do not know why they are taken and know that they have been. It is ... our goal to try to make as little impact on human psyche, human (*slowly*) *psyche* ... as possible.

Leo: Is there information about the percentage of encounters or abductions that are conducted by Greys versus other groups?

Grandpa: I do not know that. I know that I work with Stan. We work with Stan ... and that we work with others.

Leo: Do you know these other groups?

Grandpa: I know of them. I know some do not have ... the same motives. And their motives are not what you call ... (*slowly*) *positive* ...

Leo: Benevolent? So some are more abusive or hurtful?

45

Grandpa: Yes.

Leo: And do you know what the difference is between those groups?

Grandpa: Some are concerned ... with their own ... *word* ... motives. Some are concerned with ... mankind and ... what direction they are going. Mankind is at the verge of ... space exploration, but they are hostile and ... they must know that they will not be allowed out ... into ... (*gesturing outward*) if they continue to be hostile.

Leo: Is there information available through Stan about Stan's sister and her health? [Stan's sister]?

Grandpa: [Stan's sister]. We are not sure ... but she has been ill. We are not sure of the cause. It is not from us.

Leo: Not from your taking her?

Grandpa: No, it is from ... it is from ... her own race ... mankind.

Leo: Is this information accurate? Is it a way to protect knowing what happened to [Stan's sister]?

Grandpa: I do not understand.

Leo: Is this a lie? That you are taking her for your own purposes, or is it somebody else?

Grandpa: [Stan's sister] is ... because [Stan's sister] is ... *word* ... related to Stan. She is not ... the right one. Stan is the right one.

Leo: Stan is the "right one"?

Grandpa: Yes.

Leo: If this matter is urgent, is there a reason why you did not contact Stan and others earlier?

Grandpa: Urgent.

Leo: This meeting?

Grandpa: This meeting is for questions that you have.

Leo: OK. Is there information about DNA that Stan has? Has it been altered?

Grandpa: Yes. Slightly.

Leo: Is there a reason for this slight alteration?

Grandpa: To (*slowly*) *better* ... align ... to better ... *word* ... communicate, to understand ... to ... know. Stan is intelligent but communication like this would not be possible without ...

Leo: Some modifications?

Grandpa: That's correct.

Leo: Thank you. Now here's a series of questions that are presented, and we hope that you are willing to respond to them: Who are the three other people who are being given information like Stan?

Grandpa: Susan Carson. Susan.

Leo: Susan knows?

Grandpa: There are ... six others ... besides Stan.

Leo: Six others?

Grandpa: That will ... provide information. That will ... help.

Leo: So that's why Stan feels that there are seven—Susan and six others?

Grandpa: Yes. There are ... seven altogether. Susan knows, also.

Leo: And how can they contact you?

Grandpa: Contact, more contact ... Susan will set it up. Susan already knows.

Leo: OK. Are there implants in Stan?

Grandpa: Yes.

Leo: Is there information about the location or purpose?

Grandpa: To monitor ... to monitor ... *word* ... to monitor ... (*slowly*) *vital* ...

Leo: Vital signs? OK. Thank you. Would you allow yourself sometime to be filmed or photographed?

Grandpa: It ... humans are not ready.

Leo: Humans are not ready for the image or the photo?

Grandpa: Humans are ... still hostile ... and ... once the ... shift happens ... then there will be no problem. We will know when the shift happens.

Leo: Is there information about the schedule or the timing of when the shift occurs?

Grandpa: When ... there are more that believe ... the human shift will happen instantly.

Leo: Meanwhile, is it appropriate for us to ask you for some kind of demonstration?

Grandpa: It depends.

Leo: Depends on the request?

Grandpa: Yes.

Leo: Is there information available about how Stan can be helpful to you?

47

Grandpa: He is providing a communication. He is ... helping ... his drive is strong. He knows ... what his goal is.

Leo: Are there others who are waiting for Stan's information to be released?

Grandpa: Yes. And the others. Not only Stan but others, six others.

Leo: Others as well?

Grandpa: Yes.

Leo: Are there individuals who should be contacted who would help with the development of this information and the dissemination of this documentary?

Grandpa: It will be ... given to the people that it needs to get to on this planet. Nothing happens by accident. Stan knows this. Lisa knows this. You know this.

Leo: Yes. Thank you. Another series of questions, some about Stan's family: Did Stan's mother play a role in his development?

Grandpa: Yes. (*Slowly*) *Unconscious*, unconscious.

Leo: Unconsciously?

Grandpa: Yes.

Leo: Were Stan's grandparents involved?

Grandpa: No.

Leo: So the involvement and work with you did not go back beyond his parents?

Grandpa: Yes. His father ... and his mother ... his siblings. Stan was ... Stan was the right one.

Leo: Have there been hybrid children created from some of Stan's other family members?

Grandpa: Yes.

Leo: Is there information about those hybrid children?

Grandpa: One. (*Slowly*) [Stan's brother] ... [Stan's brother].

Leo: [Stan's brother]?

Grandpa: Yes. [Stan's brother] one—Stan many. [Stan's brother] one, Stan many.

Leo: Is it appropriate to ask you whether you can read the formulas that Stan has received?

Grandpa: It is ... for ... *word* ... it is to ... qualify ... Stan knows he cannot write these ... others know Stan cannot write these. It is to prove that Stan's experiences are real.

Leo: M-hmm.

Grandpa: Susan knows ... Susan has people that know of the equations mean—(*slowly*) [*name withheld*] ... [name withheld]?

Leo: And others? Do you yourself have knowledge to use the formulas?

Grandpa: I do.

Leo: Is it appropriate to ask the purpose of the equations?

Grandpa: It is ... propulsion. It is ... it is ... travel. It is ... knowledge that ... is already ... out ... but mankind needs to focus more on that direction.

Leo: Focus on propulsion and space-time travel? Is that the purpose of the equations?

Grandpa: That explains things you do not understand. Yes.

Leo: The symbols on the equations that look like pathways—can you describe the meaning of those symbols?

Grandpa: They are ... connection ... they are ... they are ... (*gesturing with thumb and forefinger*) bringing two points together. Humans have a hard time ... thinking ... in a ... four-dimensional reality, let alone five- or six-dimensional reality.

Leo: Those are many dimensions involved in these equations?

Grandpa: There are many ways to travel. You do not need to take years to travel great distances. You can ... shorten the distance (*gesturing with thumb and forefinger*).

Leo: Are there doorways or pathways in which space-time is diminished, or—

Grandpa: There are occasions ... that make it easier to get from one place to another ... you might call them ... *word* ... gates.

Leo: Gates?

Grandpa: Gates.

Leo: And is there specific information that is available regarding how these equations could be used, should be used—or is that for others to decide?

Grandpa: It is ... our intention to guide. We cannot get involved ... we are not allowed to do that ... but we can push in the right direction. Man must learn on their own. Man ... wants sovereignty; they must learn on their own in order to get sovereignty. We cannot help in this way.

Leo: Is there significance in regard to the information about planetary alignment that you've given Stan?

Grandpa: It is ... been given Stan ... the date.

Leo: That date in September of 2012?

Grandpa: It is. The date has been figured out, yes.

Leo: The events that are likely to occur at that time?

Grandpa: Hmm ... it is not important.

Leo: Not important for human understanding—at this time?

Grandpa: You must ... humans must concern themselves ...

Leo: Are there other events that will lead up to the transition?

Grandpa: Yes. But humans must concern themselves with ... enlightenment and knowledge. There is help. There is ... guidance that will be available once ... they realize that they are not alone. And ... they can prove that ... they are beyond hostility.

Leo: The important factor is human awareness of the presence of other civilizations?

Grandpa: Yes. Things will happen automatically. Things will happen ... when ... when the shift ... when the shift occurs. We will know when it happens. Stan ... and the others are there to help with that shift. So ... humans can be part of ... word ... (gesturing) ...

Leo: Part of the ... cosmic culture?

Grandpa: Yes.

Leo: Is there a title or a term used? "Galactic Federation" ... "cosmic culture'" ... "extraterrestrial civilization"?

Grandpa: It is a ... governing ... group. There is ... a governing ... word ...

Leo: Like a council or a group?

Grandpa: Yes. There's a governing council. And humans must ... prove themselves, or humans will destroy themselves.

Leo: Is there information about whether there'll be natural events as well as social events that will lead to the change or the shift?

Grandpa: The shift ... will happen ... when humans agree they are not alone. They are being ... word ... subdued. They are being ... guided incorrectly ... and ... this has caused ... problems with their development. They are ... word ... this has caused problems with their evolution ... mental and physical evolution ... humans do not understand who they really are. They are a water vessel. They are a vessel for something greater.

Leo: And when these changes occur, humans will learn more about origins?

Grandpa: It will be instantaneous. A ... like a ... (*slowly*) *switch* ... a switch. Switch. Switch.

Leo: Is there information available regarding the photon belt or galactic superwave?

Grandpa: There are many types of energies.

Leo: The energies that come through this part of the solar system?

Grandpa: It is being constantly bombarded with particles, yes.

Leo: And these energies will have some influence on humanity?

Grandpa: Hmm ... humans cannot understand ...

Leo: But there are some human scientists who are saying that is energy coming through this part of the galaxy.

Grandpa: There is energy that affects human life. There's energy that affects all life, some negative, some positive.

Leo: I'm talking in terms of earthquakes and volcanic eruptions on the planet that have an impact on the—

Grandpa: Galactic energies. There are energies that mankind does not know of yet or understand.

Leo: In the last session, you mentioned that the greedy evil rulers of Earth will be dealt with. Is there information about how they'll be dealt with?

Grandpa: It is not important.

Leo: Not important what can happen to them?

Grandpa: It is ... the message is for everyone. The message is for all; the message is for "good" and "bad." Humans do not understand who they really are, or the connection they have with each other. The ones that are evil are lost and need guidance. There are some that cannot be guided ... some that ... *can* be guided.

Leo: OK, thank you. The next question is about this film, the videotaping. Is there information known through you about who might be helpful in producing such a documentary?

Grandpa: It is ... been allowed so information can ... be given. Humans have to find their own way, but we can guide. We can point; we can guide in the right direction ... and ... it will happen when the time is right.

Leo: OK. In the previous session there were some limits to the communication in terms of time and information. Was there a reason for those limits?

Grandpa: Communication ... is ... we do not communicate as humans. Communication like this is hard for Stan and taxing for Stan so ... it is ... it is our goal to ... make little impact ... on Stan as we can. It is our goal to guide. We cannot provide ... every answer. Mankind must find it for themselves. They insist on sovereignty. They must create their worthiness ... to get sovereignty.

Leo: Thank you. Is there information about whether humans can increase [their] telepathic abilities?

Grandpa: It will happen when the shift happens. There are those even here that have that ability. And as time goes on, it will get stronger.

Leo: Are you able to provide us with a demonstration of speaking in ancient or difficult human languages?

Grandpa: It is not necessary.

Leo: Is there information about the location of your home planet or dimension?

Grandpa: We are ... by a planet ... we are ... six light years down and off from ... where you call ... (slowly) Alnitak ... Alnitak ... Alnitak. Zeta, Orion ... pyramid Khufu ... Alnitak ... we are behind Alnitak and off from your view. We are six light years behind what you call "Orion," on the other side of that. We are ... we have a view of ... horse ... you ride them ... word ... (slowly) horses head ... nebula. You cannot see it ... we have view of Horsehead Nebula.

Leo: Thank you. And your function ... is it diplomatic or academic? Do you have a title or a function?

Grandpa: We do not ... think in those terms; we are beyond ... thinking in those terms. I am a guide. I am a ... what you call ... (slowly) research ... research. But I am mostly a guide. I guide Stan and there are others like me ... to help guide.

Leo: Thank you. Now we have some questions from several people—some might be repetitious—and we'd appreciate it if you were willing to respond to them.

Grandpa: Yes.

Leo: Questions from Robert: Do you have a family?

Grandpa: We do not have families like you have. We do not procreate like you do. Others do not ... we are many species from what you call the universe. So we do not need to create life ... and mates. Some create life on their own by themselves. We are beyond that.

Leo: Is that so? Do you nourish yourself with food? Do you eliminate body waste?

Grandpa: We ... do, but not like humans do. All are different ... [unintelligible].

Leo: OK. Is there a reason why it's so difficult to make contact with humans?

Grandpa: We do not ... we cannot ... interfere. We do not want to cause fear. We do not want to ... interrupt the human experience. Human emotions can be painful.

Leo: Are there ways that we could help in terms of communication and contact?

Grandpa: Yes. The shift ... must happen when humans realize there is nothing to be afraid of. When humans realize that ... they are not alone. That will help. It will happen automatically. Humans have ... (*slowly*) *purposely* ... purposely have been ... misguided and ... their abilities have been ... (*slowly*) *subdued*.

Leo: Subdued?

Grandpa: Yes.

Leo: Stan has heard footsteps on the roof. Do you know why?

Grandpa: (*Laughing*) Stan is inquisitive. We make ... *word* ... experiments ... some ... to see what happens with Stan, sometimes for enjoyment.

Group: (*Laughter*)

Leo: Do you know about rocks that have been thrown at Stan's window?

Grandpa: Hmm ... that is not us.

Leo: That's not an experiment or a joke?

Grandpa: No, that is something else. There are ... others that are involved that are ... not us but that seem to be attracted to Stan and ... because Stan is different. He ... is bright. He is brighter than most.

53

Leo: Do you have information about the lady who visited Stan in his childhood?

Grandpa: Yes.

Leo: Is that something you can share with us?

Grandpa: Yes. Offspring.

Leo: Offspring.

Grandpa: Stan's offspring.

Leo: Stan's offspring. She … she has information about Stan's offspring? Or …

Grandpa: She is Stan's offspring.

Leo: She *is* Stan's offspring? Then how can she be with Stan when he is a child?

Grandpa: Humans do not yet understand. There are many ways to travel.

Leo: OK. And do you communicate in other ways besides what has happened with Stan … in his life? Are there other kinds of communication?

Grandpa: There are many forms of communication throughout what you call … the universe. We do not communicate like humans. We must utilize technology … human technology to communicate with humans. And or … utilize Stan to communicate with humans … or others like Stan.

Leo: Are you able to disrupt gravity when standing on our Earth?

Grandpa: Our technology is advanced … more advanced than humans'— but there are others that are more advanced than we are. There are others that do not need physical forms. In fact, humans only use this existence with physical forms, but humans seem to be more … we are learning.

Leo: Do you have any capacity to help Stan's body float now?

Grandpa: It would take … it would take energy that is not allowed … and it is not necessary.

Leo: On the transition, does the Earth 2012 … will humans evolve to another dimension?

Grandpa: We are hoping that … the trans … the transition of knowledge will happen sooner. There are different types of transition. If the transition of knowledge does not happen sooner, humans … will face peril. Humans will face … possible extinction.

Leo: If humans are destroyed or extinct, will the soul continue on its journey?

Grandpa: We are learning. But it seems to be ... humans are more than they understand. Humans ... are being studied. Humans were designed by ... beings ... more advanced, as we are more advanced to you.

Leo: You are learning from those that made humans?

Grandpa: We are learning. Those entities are beyond physical form. We know of them, but they are beyond our comprehension—*not* comprehension—they are beyond physical form.

Leo: OK. Do you know if Stan is tiring? Is it appropriate to ask you to come back at a later time?

Grandpa: Yes.

Leo: About ten minutes?

Grandpa: Yes.

Leo: OK, thank you very much for being here. We really appreciate the information you've given. We will ask you to return. Thank you.

Dr. Sprinkle returned me to an awakened, conscious state. After telling me that we are taking a break, I mix with those attending to get a sense of how everyone is reacting to the information. Once I feel my energy return, I let Leo know that I am ready to continue with the session. Leo begins the session again.

Leo: Grandpa, are you with us again?

Grandpa: Yes.

Leo: Thank you so much. We appreciate your willingness to respond to many questions. We have some questions that may seem repetitious but appreciate your willingness to respond. Is there information about the interpretation of the date, 1 ... 10 ... 6, 4, cheese, 9, 8 ... any information about that date? [We discovered later that "cheese" was a humorous reference to our moon.]

Grandpa: It is ... an alignment. This alignment will happen approximately ... when the ... this was the date given to Stan. Stan was curious as to when ... Earth changes will transpire ... and this date was given to

55

Stan. It is alignment of planets. Stan did not understand, and this was a way for Stan to understand when the date was.

Leo: And earlier information about that date now?

Grandpa: Hmm … it is not important.

Leo: All right, thank you. Is there information about cryptology or writings, which are given within the equations?

Grandpa: Stan is given the equations as validation so others know that his experiences are real. There is some information to help guide and … (slowly) push … push … push...push mankind into the correct direction.

Leo: Are you or the group you work with, are there races who are working with the U.S. government?

Grandpa: There have been, yes. And there are now, yes.

Leo: Is there information about the Starseed letter? Where did that come from?

Grandpa: (Slowly) Starseed letter … it is a name, "Starseed letter"? Explain.

Leo: Sarah has information about a letter that was called "Starseed—Hello, Starseed." It's difficult to read the front page, but the second page has to do with information about a program: "Three-dimensional reality reaches to end in one level of reality into the beginning of the next. Our species seems to be at a plateau now." So apparently this is from an extraterrestrial source. Is there any information?

Grandpa: There is … truth in this. There is … that is a … human … understanding of a reality that is … happening at this moment. There is … levels that every civilization must take. There are many civilizations in many levels throughout what you call the universe. Some are … advanced; some are as you are. There are many.

Leo: Thank you. The cryptology question had to do with the symbols that appeared from the writings that Stan did. The equations. And then at the bottom there were some symbols.

Grandpa: It is validation that Stan is going through … what Stan is going through … they are ancient writings. They are things that Stan will have had no knowledge of, to validate Stan's experience. Stan must have validations for other[s] to believe that this is real, and that information can (slowly) progress, pro … progress, progress … that information can be given to those that need the information.

Humans do not ... humans are still unaware. They do not believe that there [are] others out in ... *word* ... universe, out in the universe beside them and ... they must learn that there is more ... than just them.

Leo: Thank you. Is there information available regarding the laws of the universe? Are there universal laws and rules?

Grandpa: There are laws within laws. There are ... more just *knowing* ... that certain things are to be done and certain things are not to be done.

Leo: Are there consequences for those extraterrestrial or those groups who break the rules?

Grandpa: There are known consequences of *any* action. Humans know this, too, you do not ... *word* ... violence ... *word* ... violence begets violence.

Leo: Is that the consequence for engaging in violent action?

Grandpa: That is correct. Most intelligent life forms understand this and ... by nature are not. The more intelligent they are, the less violent they are. They know that the ... (*slowly*) *reper* ... *repercussions* ... they know the repercussions, and only primitive planets are ... violent. There are some, however ... there are some intelligent planets that are ... aggressive and ... it is in their nature to be so. They do not care if man succeeds or not. These are ... *word* ... *word* ... they look like ... represent dragons ... *word* ... lizards ... dragons ...

Leo: Reptilians? Lizards or—

Grandpa: There are creatures that resemble this, yes.

Leo: Thank you. Can you elaborate on the planet alignment? Can you talk about ... is information available about what happens?

Grandpa: (*Deep breath*) It is not necessary.

Leo: OK. Is there information on the history of Mars and [its] connection with Earth?

Grandpa: Mars was ... Mars was a ... thriving ecosystem, but ... Mars was too small and could not support ... the ecosystem. Mars had a ... and *still* has a small ... *word* ... um ... to monitor, to monitor humans ... monitor humans. The moon also has a ... *word* ... monitor humans ...

Leo: A base or ...

Grandpa: Yes. Base. Thank you. Humans have been to this base.

Leo: Thank you. Sarah wonders is she can request an object be brought here from another planet, another dimension, so that she can study it. Is that possible?

Grandpa: When humans are ready, it will happen, but it is not … it is not time; it is not necessary.

Leo: Is information about the base map created by a man named Ivan Sanderson, regarding undersea objects, submerged objects …

Grandpa: There are things on this planet. There are bases on … the planet you call the moon. There are bases on Mars. There are … ships … (slowly) ships … ships … (slowly) nurture … (gestures as though holding an infant) … word …

Leo: "Mother" ships?

Grandpa: *Mother* ships! Mother ships … in various parts of your system.

Leo: Are there some beings that we should be more cautious of, or some we should pay more attention to?

Grandpa: Hmm … it is good to … question. It is good to be (slowly) suspicious, but it is also good to have an (slowly) open … open mind … open mind. There [are] those out there, like there is here … some do not want mankind to succeed and some do. And … at this point we are hoping that mankind will realize they are not the only ones. That they are … it is imperative; it is (slowly) imperative that man knows they are not alone. The shift will happen when this happens. We will know when the shift happens.

Leo: Thank you. Do you have information on the life of Jesus?

Grandpa: Jesus … (slowly) Jesus. There are many … religious … there are many religions, which … word … were … put in place to control man … to show man right from wrong. Primitive man … word … man is hostile … man … has been led incorrectly. It was hoped that man would learn that … this is not correct. There *is* … there *is* God … there is … you call "God" … there is … God. We do not understand, but there is something more. And … the universe is … connected. All the universe is connected … word … man cannot understand. Man does not have the ability to … understand. The concepts are beyond man. And … there is more … there is more, there is a connection with everything and … man must first … man must takes steps. Man must … understand they are not alone.

Leo: Are there other civilizations or groups in this solar system?

Grandpa: There is life ... not what you call intelligent, but there is life on other bodies in this system. Intelligent life is ... intelligent life is ... based throughout this system. Based ... not *from* this system, but ... based here ... to monitor man. There was at one point another intelligent race on a planet that no longer exists. What is left of the planet is ... (*slowly*) Mars ... past Mars ... Mars ... (*gesturing*).

Leo: The asteroid belt?

Grandpa: *Asteroid belt!* Asteroid belt.

Leo: Thank you. Here are some questions from Thomas: The flashes or light or the orbs, is this a technological phenomenon, or is it a natural phenomenon?

Grandpa: Some, as I discussed, some orbs ... what you call orbs ... orbs ... are not from us. Some are ... flashes are from us. Flashes are monitoring. Flashes are ... *word* ... step through ... step through ...

Leo: Gateway?

Grandpa: Doorways.

Leo: Doorways? From other dimensions or ...

Grandpa: Um ... humans do not understand.

Leo: OK. How about your own life—yours and those around you: Are you more biological or more mechanical?

Grandpa: We are ... biological. There are many different types of life. There are ... some life that is ... manufactured. There is life that is beyond us ... the more advanced life is, the less physical they are. Man has ... we are studying man. Man seems to have both. Man is a water vessel containing something else.

Leo: Thank you. Are you able to describe your home? Your planet, in particular your own individual housing or home?

Grandpa: We are ... we are not like humans. We are ... we are many places.

Leo: You are able to be in many places?

Grandpa: We are ... in many places. We are not just located like humans are, confined to this planet. We are many places, and where environment dictates our living arrangements, we are ... we modify our living arrangements to the needs of ... or the environment [where] we are staying at the time.

Leo: Thank you. Do you know if any humans on Earth possess some of your technology?

Grandpa: Yes. (Slowly) government ... government. There are a few. This was supposed to be shared, but it is not being shared.

Leo: I see. Is that one of the concerns about what will happen at the shift?

Grandpa: Yes. The shift will ... we will know when the shift happens. This ... some technology, what you call technology, has been shared; some has not. Humans will ... create their own doom, their own destruction because ... they are using incorrect forms of transportation. They are damaging their environment. They are ... hurting and killing themselves because of greed and because of ... word ... because of ... selfish gain. And ... this technology, this was meant to enlighten, and it has not been shared. Some has, but some has not.

Leo: Thank you. Do you have information about the female whom Stan met ... on a craft? Her name or any information about her?

Grandpa: That is not important, but it is Stan's ... (slowly) offspring, offspring.

Leo: Thank you. There's some questions from Mark: Many of us have given our lives or have dedicated ourselves to getting out the truth of ETs' existence or exposing corruption. Can we expect our efforts to be rewarded or unrewarded?

Grandpa: Yes. They will be rewarded in ... word ... there is gratitude and ... they will be rewarded when the shift happens.

Leo: Thank you. Are you aware of information about afterlife or spiritual world?

Grandpa: There [are] many forms of life. Some we understand, some we do not. There is ... those that are more advanced, as we are advanced to you. And ... there [are] ... different levels of existence ... beyond even our understanding. There is a ... word ... there are ... many ... levels. There are many ... what you call (slowly) dimensions, dimensions.

Leo: Many dimensions? Have you been thought [of] in the past—or those of you that have come to Earth—been thought of as gods by humans?

Grandpa: Yes, some have. Some contact ... there has been situations in which we have been considered ... deities or "gods," like you said.

Leo: Do you have knowledge about the future of humankind?

Grandpa: Hmm ... future is ... hard. Future is ... past is easy. Future is ... humans do not understand ... *word* ... future is changing, moving, changing ... cannot, cannot pinpoint ... one small movement may affect what happens. Everything is connected in one point. Future, past, present are all connected. Dimensions are all connected with every other, every other dimension. Humans do not yet understand. It is beyond their comprehension.

Leo: Is it likely that humans will become vegetarians, in the long run?

Grandpa: I do not know.

Leo: Is there information about whether you or your colleagues can be called upon by those people in this room?

Grandpa: If it is necessary we will ... communicate. This is a communication for your purposes.

Leo: Mark wants to know if there is anything that can be done—can individuals here help more in your work?

Grandpa: Yes, they must ... focus. They must ... help Stan and Lisa. They must ... let others know that it is not meant to be fearful; that it is for enlightenment; that there [are] others out there besides themselves. There is a ... a con ... a ... (*slowly*) *concerted* ... a concerted effort to ... push the consciousness forward. Everything has been ... *word* ... everything has been (*slowly*) *carefully planned* ... and everything is working for this goal.

Leo: Thank you. So each person here has duties or tasks?

Grandpa: Yes.

Leo: Some people want to know what they can do to be more helpful.

Grandpa: They are doing what they are supposed to do. Everyone here has been chosen. Everyone here is not here by mistake. It is ... they are known to us who they are. They are ... they are ... to strive to ... they must explain that there is more ... they must explain to others that there is more out there, and in order for the shift to happen ... humans need to know that they are not alone. Humans ... when the percentage of humans in this world is greater that know they are not alone, the shift will be automatic, and the rest will follow.

Leo: Thank you. A question about Susan Carson: Is she one of the special group, or does she just know of —

Grandpa: She knows of the seven. She is … *word* … she is not the seven, but she is … the … lead … in the group … she will … *word* … bring them together. It is important that they be brought together. To understand that Stan and the rest must be brought together. It is … it will be enlightening.

Leo: Thank you. A question about specifics in regard to DNA: Is there information about strands, genes, sequence … is there any important information from Stan's genetic background?

Grandpa: Stan is different. There is … subtle variations that have been purposely put in Stan's what you call DNA. And there are … there is evidence of this DNA in every human. It is dormant. It will awaken when the shift happens.

Leo: I think you answered a question like this before, but the question is, are you able to perceived the future, and if so, how do you do it?

Grandpa: Hmm … future is constantly changing. It is hard to … it is hard to predict; it is hard to … know what the future … brings or what the future holds because it is constantly changing. The smallest … thing could switch directions, and the future is not like the past. The past is … preset; the future is not. The smallest thing could cause the future to … redirect itself.

Leo: Thank you. This question is from Paul: Does your group or race believe in a divine creator?

Grandpa: Yes. Some do not; some do. But there is evidence that there is something more than us. We know that it is there, but we … do not understand it. There seems to be a oneness with everything, a … as if … fragmented, *as if* … this oneness was fragmented and branched off. There is a oneness.

Leo: Do your people perceive the presence of the spirit of deceased beings?

Grandpa: There is no "deceased." There is just change.

Leo: So the soul continues on?

Grandpa: It is apparent to us that it does, yes.

Leo: This question's from Robert: You mentioned that you were not the ones throwing rocks at Stan's window. Do you know who is throwing the rocks, and why?

Grandpa: There are ... there are ... others, not us. There are ... malevolent ... they are not threatening; they are ...

Leo: More mischievous?

Grandpa: Yes, (slowly) mischievous.

Leo: Thank you. Do you know who is making candlesticks fly off the shelf at Stan's house?

Grandpa: (Smiling) ... yes.

Leo: Can you describe them?

Grandpa: It is not important.

Leo: Same about the chairs placed on Stan's roof?

Grandpa: (Smiling) Hmm ... yes.

Leo: Mischievous?

Grandpa: Yes.

Leo: Thank you. You mentioned you don't reproduce like humans. How does your species reproduce?

Grandpa: It is ... a process. We have ... developed beyond the need for interaction ... with others of our kind. We can produce offspring when we need. We do not ... haphazardly overpopulate and destroy the environment. We produce, as we need.

Leo: Thank you. A question about the flashes of light that appeared outside: Are they from you or your group?

Grandpa: One.

Leo: One?

Grandpa: One.

Leo: One person or one occurrence?

Grandpa: One flash.

Leo: One flash. Is it possible to manifest an orb or light in this room?

Grandpa: It is not needed.

Leo: Thank you. OK, now we have other questions, if you are willing, from the group, Alejandro and others.

Grandpa: Yes. OK, proceed. Wait ... (raising finger for a moment, then lowering finger). OK, proceed.

Leo: Thank you. Is there information about free energy devices? Do you have knowledge of people who have free energy devices?

Grandpa: There are humans that ... know. There are humans that understand that—um ... they are being subdued; they are being held back.

Leo: (*To Alejandro*) Yes, Alejandro, I have a VHS and a DVD about a free energy device that's called *Race to Zero Point*. Be glad to share it with you. (*To Grandpa*) Are there certain groups of religions or these systems that seem most accurate to you and your group ... human religions or human philosophies?

Grandpa: They are meant to ... guide. They are meant to ... push humans to understand that ... there are consequences to violence; there are consequences to ... evil intentions. There is a ... universal understanding that love is ... the correct way. Love is ... the way to proceed ... to ... *word* ... a ... government ... divided ... will fall.

Leo: Are you familiar with some of the humans who have comprehended these truths?

Grandpa: There are ... a few, yes. Jesus ... Gandhi ... one you call (*slowly*) *Martin* ... Luther.

Leo: Martin Luther King? Thank you. Has this kind of transition that's going on here now, has this happened in other civilizations as well?

Grandpa: Yes, throughout the universe. Some have made it; some have not. We hope that humans ... can ... grow up.

Leo: If we were to use a rating scale between 100 and 1, 100 being the most advanced civilization, at what level would yours be?

Grandpa: Hmm ... numbers ... *word* ... we would be ... thirty, forty percent; we will be forty. Um ... maybe not forty, maybe thirty.

Leo: Thirty to forty? Thank you. When you say there are many interested in our advancement, do you know how many civilizations are interested in human advancement?

Grandpa: There are ... it is not important.

Leo: Are there different origins of the various races on Earth? Have we come from different dimensions or different planets?

Grandpa: Humans are ... from many different species. Humans are a conglomeration. Humans are of this planet, of others. Humans are ... humans seem to have been designed by ... more advanced than us. More advanced as we are to you; they are to us.

64

Leo: I see. Thank you. Is there a litmus test, or is there a way of knowing when an encounter or an abduction is being conducted by a malevolent or harmful group?

Grandpa: Humans have not completely lost their ability. They know when something is not right. They instinctively have that ability, when to be cautious. It is good to be cautious. It is good to be cautious without being paranoid.

Leo: Cautious without being paranoid. Thank you.

Grandpa: (Looking directly at Paul) Paranoia is not good. Paranoia is bad. There are those in this room that have taken it beyond a healthy level. Paranoia is not good.

Leo: Thank you. Do you know if there are some humans on this planet that have developed the technology to travel through time?

Grandpa: No.

Leo: Are there some secret groups who are working on this path?

Grandpa: Yes. They can ... man does not understand what power they have. These men do not understand what power they have. They are dangerous. They are ... a ... controlling entity. They are controlled ... they're not ET; they are human; they are controlled. They are a controlled group that seeks power; they ... stop at nothing to get power. And they want to resist for their own purposes. We are closely monitoring their ... progress and ... we will intervene if necessary, but ... this knowledge should be shared with all humans, but there is a problem. There are some humans that are very hostile and will use this for the wrong purpose.

Leo: Thank you. Are you able to visualize people in this room through Stan's body?

Grandpa: Yes. Why?

Leo: Robert wants to know ... (awkward laughter) are you able to see us or visualize us?

Grandpa: I do not see like you see.

Leo: I see. (Laughter) Thank you. Is there information from your group about the journey of the soul, what happens to it when the body dies? When the soul moves on?

Grandpa: We are studying it; we do not yet understand it.

Leo: Do you have information about what humans may do with additional senses or additional abilities?

Grandpa: They will join the ... Stan calls it "the neighborhood."

Leo: The neighborhood. Yes. Thank you. Do you have information about how many different dimensions are foreseen or understood in the world?

Grandpa: There are more than three.

Leo: (*Laughter*) Do you have information about doorways to other parts of space, like wormholes ... gates?

Grandpa: Yes. Yes, and so do some humans. They are ... we are trying to ... push humans in that direction. Humans have been taught incorrectly. There is a more efficient way of travel. They say that ... human scientists say that the fastest way from one point to another is a straight line. That is incorrect. You can touch ... touch ... (*gesturing*) bring them together.

Leo: Different parts of space, of the space-time continuum—can bring them together?

Grandpa: "Continuum." Hmm ... space. Bending ... space.

Leo: (*Laughter*) Thank you. Are you able to identify people who can levitate or teleport?

Grandpa: Not here.

Leo: Not here? Not here on the planet?

Grandpa: Hmm ... there are some that have succeeded in ... some minor, some small ... abilities ... but not like the more advanced civilizations.

Leo: Now these more advanced civilizations use technical as well as—

Grandpa: Yes. There are different ways of travel. There are ... some that can ... by thought; there are some that you ... can manipulate. There are chemical, spiritual—some you call spiritual—a combination.

Leo: Combination. Thank you. Do you have a testament of the percentage of humans who must evolve to a certain point for the shift?

Grandpa: "Evolve" will happen naturally when ... a percentage of the world, Earth—of your Earth ... of humans—believe that they are not alone—that accepts that there is life other than themselves. Right now it is ... twenty-five to thirty percent, maybe more.

Leo: When it gets to ... the middle fifty percent ...

Grandpa: When it gets to ... when it gets past what you call "half, half, half-way."

Leo: That's when the shift will occur?

Grandpa: Yes, that's when the shift will begin. The more, the faster they believe, and the more people that believe, the faster the shift. And the shift will gain momentum.

Leo: I see. Thank you. Is your group involved in cattle mutilations?

Grandpa: No. This group is not, no.

Leo: Do you know which groups *are* involved?

Grandpa: No, I do not know.

Leo: Other questions, if you are willing: Are you and your group able to make objects move mentally?

Grandpa: If need be.

Leo: Such as making a candle fly across the room?

Grandpa: Yes. (*Smiling*)

Leo: Is that what the group usually does?

Grandpa: Um ...

Leo: The smile on Stan's face says it must be ...

Grandpa: It depends.

Leo: There's no purpose to the situation?

Grandpa: That is correct.

Leo: Is there information from your sources available about cattle mutilations? Ghosts? Sasquatch? Pleiadians? Reptilians? Chupacabra?

Grandpa: Sasquatch ... there are many different species in the universe; there are many. I am not familiar with Sasquatch.

Leo: Sasquatch refers to "Big Foot," a primitive, large creature.

Grandpa: There are ... some species that will be discovered. I believe this is what you're talking about. They are intelligent but in a different way.

Leo: Do you have a range of emotions? We know you have a sense of humor ...

Grandpa: Our emotions are different. We do not dwell on emotions. Humans dwell on emotions; they are guided by emotions. They must grow up. Emotions are good to an extent, but they cloud judgment ... and we do not have time to have clouded judgment.

67

Leo: Thank you.

Grandpa: (*Raising a finger*) Stan is getting fatigued.

Leo: OK. Is it time to end the session?

Grandpa: Hmm ... a break.

Leo: Thank you so much for sharing information. Thank you.

A break is taken, after which the session continues. Leo induces me back into a hypnotic state.

Leo: Thank you. Thank you for being here, and thank you for responding to questions. A question has been written by Robert: Do you know why Lisa has a triangular mark on her head?

Grandpa: Yes. It was ... to monitor ... it is a ... to monitor.

Leo: To monitor Lisa's behavior?

Grandpa: Hmm ...

Leo: Or her ... reactions?

Grandpa: Yes ... and no. To monitor ...

Leo: Vital signs?

Grandpa: Yes, vital signs and ... other things.

Leo: Emotions or ...

Grandpa: Yes, that too.

Leo: Thank you. Do you have information about the crop circles—how they're formed, and why?

Grandpa: Crop circles are ... not all ... but it serves the same purpose. Crop circles are to push mankind ... subconsciously ... and consciously, in the direction they need to go. They will ... they create ... curiosity. Curiosity that ... will ... *word* ... will bring out ... *word* ... will push man to ... seek the truth. There are ... crop circles that are ... *you* call them crop circles; we do not call them crop circles. There are impressions that are made by us, and there are impressions that are made by man. But they both ... serve ... the same purpose.

Leo: Increasing understanding?

Grandpa: Exactly. They are ... designed to ... (*slowly*) spark ... *word* ... curiosity ... so man will be led in the correct direction.

Leo: Are they a combination of ancient knowledge and what is current technology?

Grandpa: Hmm ... this has been tried before ... in early development of man, but you were not ready. They considered them ... of evil ... *word* ... not to aliens or to do with religion—

Leo: Evil spirits were attached to them or beings of satanic—

Grandpa: Yes.

Leo: And does your group work hard to create the crop circles?

Grandpa: Not my group directly, no ... but others—like us and other races. There are ... some that ... just work particularly on ... Stan's word ... particularly on ... that, but we ... have our own job to do.

Leo: Has your group been part of the establishment in the building of the pyramids?

Grandpa: Hmm ... no. Man does not give himself enough credit for ... some of the structures that have been built. We have guided, but we ... did not build the pyramids. No. And Susan must know that ... we have shown her; this must get to Susan. We have shown her ... how we can make an impression. She has seen this, and it is a gift for her.

Leo: Thank you. Now, some questions from Paul: Were the Apollo moon landings faked?

Grandpa: Hmm ...

Leo: Did humans actually go to the moon?

Grandpa: Yes and no. There were some ... there were some ... some falsities and some things that happened. The first mission was successful, the other ones, some of the other ones were not.

Leo: Some people have claimed that the moon and Mars are now off-limits to human exploration? Do you know if—

Grandpa: They are now, yes.

Leo: They are off-limits?

Grandpa: Yes. They are hostile. They are not allowed to ... until they grow up. We will allow a little bit but ... we are monitoring them. When they are able to control themselves, then we will allow them to explore their own ... *word* ... area ... *word* ... sun ... solar ... (*slowly*) *solar* ... solar system.

Leo: Is one of the rules "No weapons in space"?

Grandpa: We are monitoring them. They do not understand that …
their reactions … their … violence … resonated throughout different
dimensions when they drop … atoms … drop atoms …

Leo: Atomic bombs?

Grandpa: Yes. When they drop atomic bombs, it has repercussions
greater than they can possibly understand.

Leo: Thank you. Now here are some questions from the group, asking
questions about your own awareness: Do you have more senses
than humans have for you to perceive the world?

Grandpa: Different senses. We do not talk like they talk. We do not
hear like you do. We … our communication is … the senses they do
that … they have, but it has been turned off. They do not perceive
the sense that we have.

Leo: Thank you. In terms of human years, or Earth years, can you tell us
how long your people live?

Grandpa: Hmm … hard to explain … we do not … it is different for us.
We … we … live longer … we live longer lifespan, but we do not
look at things as you do. We are multidimensional and … the way
you look at time is different. We are older but we … perceive time
differently.

Leo: Thank you. Do you have an estimate of the percentage of beings
walking around our Earth who are not actually human?

Grandpa: It is not a large percent, but there are some. There are …
some that Lisa and Stan know that are not human … they are there
to monitor.

Leo: OK. Do you have information about the numbers and places of
the space-time portals in Colorado, in this region?

Grandpa: There are … portals, strange portals … gates. There are gates.
When you see a flash in the night sky … sometimes it is … a gate.
Stan could see a gate … in the western … western, western …
sunset … western …

Leo: Western part of the state?

Grandpa: Western part of the sky. When he looked west, he could see a
flash that is a gate.

Leo: Do you have information about the number of gates in this area?

Grandpa: It is not important.

Leo: OK. Is there information about underground bases?

Grandpa: It is not important. It is not important.

Leo: Is there information about whether there ... go ahead. (*Grandpa raises a finger as though asking for a pause. In a moment, the finger is lowered.*)

Leo: Is there information about groups who may be malevolent or evil, who have agreements with the Reptilians?

Grandpa: There are some. There are some that have misguided for their own agendas, but they are watched closely. There are ... some in the government that know that they are of what you call evil intentions, and they are...using it for their own evil intentions, so ... there is evil up there, as there is down here; there is good down here, as there is up there.

Leo: It is likely to change here on the planet when the shift occurs?

Grandpa: It will change. Yes. It is up to humans to change it, though. They must ... they want sovereignty; they must change it themselves. We will know when the shift happens.

Leo: Thank you. Do you and your group hope that humans will learn to use space-time and use the propulsion information that Stan has given?

Grandpa: They will when they are ready. They are not ready yet. They will ... naturally ... when the shift happens, they will naturally ... be able to ... accept. They must first ... (*gesturing*) small steps, small steps. They must first ... accept that they are not alone before they can ... reach out to (*gesturing outward*) what awaits them.

Leo: Thank you. In your opinion, is the population of the Earth too large for the size of the Earth?

Grandpa: Very much so. Yes, that it is a concern. It is a concern.

Leo: Do you have information about the Roswell crashes, were they of your group as well?

Grandpa: We were not, but others were. Yes, Roswell is ... real. It is being ... *word* ... it is ... being subdued. The information is being subdued.

Leo: Do you know of [name withheld]? Do you know that name?

Grandpa: We do not.

Leo: OK. And your gender, do you have gender, male/female gender?

Grandpa: Hmm ... you do not have gender like we have. We've evolved past that. We have gender but they are stagnant. It is not like you have gender.

Leo: Do you have information about the asteroid belt, what happened to the planet that once was in that circuit?

Grandpa: Yes. They were a race that ... much like yourself ... was evolving but destroyed themselves because they could not ... become enlightened.

Leo: I think you responded to this question before, but is it possible that before the shift—or will it happen after the shift—that there'll be possible meetings face to face with you and your group?

Grandpa: If it is necessary there will be a ... if need be, there will be ... many. Many meetings ... if it is necessary.

Leo: Do you have information about why a man named David was called to the meeting?

Grandpa: (*Grandpa adjusts the reclining position of the chair several times before answering.*) David. Yes, he is known to us. He is ... he was invited.

Leo: Do you have anything else that you'd like to tell Susan?

Grandpa: She is important. She is on the right track. She knows. We have provided proof to her. She has been given a gift to see what [to] do. She is on the right track.

Leo: Is it allowable now that people in the group could ask you questions individually?

Grandpa: Yes.

Leo: OK. (*To the group*) Does anybody have a question they'd like to ask Grandpa?

Robert Morgan (RM): I do. Since you speak telepathically, can I ask you this through my mind?

Grandpa: Hmm ... it will be hard for us in such long distances but ...

RM: Can we try?

Grandpa: Yes.

RM: (*Silence*)

Grandpa: You do have one concern about ... I must address this ... one concern about Stan and ... (*slowly*) *funding* ... your movie ... for movies ... funding ... funding. We cannot get involved with funding.

That is up to you. We cannot help in that way. We can guide and we can assist, but we cannot help with funding.

RM: That's not what I asked. Thank you.

Leo: Anything else that you wish to say to Robert?

Grandpa: No. It is hard for us, such long distances. Stan is ... been programmed, not programmed ... Stan is ... he is a ... portal ... a communication ...

Leo: A messenger?

Grandpa: Hmm ... we can't communicate through Stan. It is hard for us to ... perceive ... we have to work hard to perceive primitive thoughts. And ... we would have to be closer.

Leo: OK. Are you willing to take another question?

Grandpa: Yes.

Leo: OK. Does anyone have a question you wish to address?

Thomas: Is there any information available on the name of the world referred to that is six light-years away from Litok [meaning to say "Alnitak"]?

Grandpa: It is not important. Susan knows it. Susan and ... researchers should know ...

Thomas: Is this the planet of origin of the entire race or just your group?

Grandpa: Hmm ... our ... our particular race has been started as ... it is past ... it is by ... *word* ... An ... Anita ... Anita ... *word* ... Zeta ... *word* ... Zeta Orionis ... Pyramid Khufu ... *word* ... (*slowly*) Anitak. It is ... you cannot see our star from ... your position without use of a mechanical device. It is ... six light-years past ... (*slowly*) Alnitak, Alnitak ... Zeta Orionis ... Pyramid Khufu.

Thomas: What [does] Pyramid Khufu have to do ... what is the connection with Pyramid Khufu?

Grandpa: It is aligned with Orion; it is aligned with the belt of Orion.

Thomas: Was that guided ... by your people?

Grandpa: Yes. Hmm ... it has shifted some in the years, but it is very close, though.

Alexandro Rojas (AR): How old are those pyramids?

Grandpa: Older than what humans believe them to be ... thousands of years older.

AR: What was the purpose for guiding that construction that way?

Grandpa: To enlighten.

AR: For us now?

Grandpa: Not so much now. More important things ... are of concern than primitive man; you are not so primitive now.

AR: Was there a civilization on Mars?

Grandpa: Not so much a civilization ... but a ... an ... ecosystem. There were bases there; there are bases now ... but an ecosystem could not support itself.

AR: Is there information available about the so-called face or the pyramids on Mars?

Grandpa: Um ... it is not important. You will discover ... mankind will discover that later.

Sarah Daniels (SD): Grandpa Grey? My name is Sarah. Do you know who I am?

Grandpa: We know of you.

SD: I experienced something ... what you would call "phenomenal" last year. Are you aware of that experience?

Grandpa: It was not my group; it was a different group.

SD: OK. What is my significance to this case, and are my experiences affiliated to the enlightenment?

Grandpa: Hmm ... you will be a messenger like ... everyone else. You are ... chosen ... in a way you do not understand yet.

SD: Am I safe to discuss this with everyone?

Grandpa: Yes. And you will have another experience, but do not be afraid.

AR: Grandpa, do you stop atom bomb explosions, ever?

Grandpa: We have, yes. Not the explosions, but we have stopped the testing. We have ... used our capabilities to show that they are not going in the right direction.

AR: Will you allow further explosions?

Grandpa: We cannot get involved; we can guide. Only if there is more at stake can we get involved. We are trying to ... my group ... our group ... trying to help ... guide ... in the right direction.

AR: So if, unfortunately, there is a nuclear war, we could destroy ourselves and you would not intervene?

Grandpa: Interesting ... we would ... never again ... if, at this point, that all is good, we could not intervene, no ... but if there were other repercussions, (*slowly*) *repercussions* ... repercussions ... there were other repercussions ... humans do not understand that what they do ... affects ... affects other things that they are not aware of.

AR: Many of us in this group and many people around Stan have started to—while we're with Stan—see ships or objects in the sky flying around.

Grandpa: Yes.

AR: Is that your group and are you ...

Grandpa: Yes.

AR: Why are you doing that?

Grandpa: For confirmation, and it will get more intense if needed. We must be careful. There are ... those here, there are humans here that are monitoring Stan and will try to ... *word* ... um ... *word* ...

Leo: Suppress?

Grandpa: Suppress, yes, but ... will try to ... stop us from ... will try to use weapons.

AR: Are some humans capable of damaging your ship?

Grandpa: No, there are weapons that humans have that most humans are not aware of ... that are ... taken ... *word* ...

SD: Biological?

Grandpa: Not biological. They are ... from us.

AR: Energy weapons?

Grandpa: They are ... *word* ...

Thomas: Reverse engineered?

Grandpa: Yes. That will work, yes.

AR: Are you mischievous at times?

Grandpa: (*Smiling*) Hmm ...

Group: (*Laughter*)

Grandpa: Hmm ... it is not important.

Group: (*Laughter*)

AR: I think you answered with your smile. Why is the cat's "meow" so significant or interesting to you? [This is in reference to Grandpa looking around when the cat walked around the chair meowing during the session.]

75

Grandpa: Humans have an ability that is rare. They have an ability to ... relate with their animals, their animals they call pets. They can understand and ... they have a gift for ... where others do not. They have a gift for ... it is a type of ... how we communicate but different ... within ... seek pets as friends ... that is unusual.

AR: Thank you very much for your assistance. Is there anything else you'd like to tell me?

Grandpa: Continue what you are doing. You are on the right track.

AR: Thank you.

Heidi (H): Grandpa, you said that some of our senses have been turned off?

Grandpa: Yes.

H: By whom, and when?

Grandpa: We are not quite sure. We are investigating steadily, but we believe that when the shift happens, it will be ... *word* ... it will come back. And ... with practice ... everybody has some ability, some more so than others. There are people in this room that have more ability than others ... in different ways.

H: And, is there anything you can tell me that I can do to help more?

Grandpa: Yes, you can ... support, you can befriend, you can believe, and help others to believe. Everyone is brought here, not by mistake but for a reason. Stan and Lisa [are] fighting a battle that is very hard for them. There are those humans that do not want this to happen because of ... selfish reasons and for their own gain. And without ... *word* ... a support system, without a good support system, Stan and Lisa could not do what they are meant to do.

Paul Wagner [pseudonym] (PW): Grandpa Grey, can you tell us why Stan's eyes appear to have elongated pupils in some photographs?

It should be noted again that all of us were under the impression that Grandpa was the "Grandpa Grey" in the video I had recorded. He is not. We would find out years later that Grandpa is, indeed, an Orion. Why he did not correct Paul indicated to us later that the Orion's place little importance on self-identity. That continues to be the case as we discover more and more about their orientation around group consciousness.

Grandpa: It is a byproduct of ... genetic manipulation, of our trying to ... tune ... Stan. It should not happen very much anymore.

PW: Thank you.

Rick Nelson (RN): How old is mankind ... humans? How old are they?

Grandpa: They are ... *word* ... you would not understand. They are forever. They are ... humans, they are forever. They are ... um ... star [cloven]? Hmm ... man ... physical man has been here thousands and thousands, almost millions of years.

RN: Was there an Atlantis?

Grandpa: Hmm ... I am not sure. I am not sure.

John: Good evening, Grandpa. Do you extern[alize] what we would term "collective consciousness" among yourselves?

Grandpa: Yes.

John: Do you know each other's thoughts? Do you experience as a group?

Grandpa: Yes. Most advanced intelligence has developed that. It is ... a collective conscience. It is a way of communicating. We do not ... communicate like you do. It is too slow.

John: If you found yourself among us, incarnated as a human being in this room this evening, how would you react? What would you do to most effectively bring information you are sharing with us? How would you guide us to share that information? What things would be most effective? What path would be taken?

Grandpa: Humans must focus. If I were to tell anything to ... people in this room, you must learn to ... *word* ... humans think ... linearly. They cannot ... think outside their line of thought. They must learn, everybody in this room must learn to think ... in a different direction ... off the ... Stan's word ... beaten path ... beaten path.

John: Outside the box? Is that—

Grandpa: What does this mean? Once again, please?

John: Thinking differently. Outside of the normal parameters that we usually do.

Grandpa: Yes, outside of (*slowly*) *the box.*

John: Are you familiar at all with the sphere of light that interacted with me?

Grandpa: Yes.

77

Mark Stahl (MS): Are you aware of what entities were behind that communication?

Grandpa: It was not us. It was a more advanced race.

MS: Are you familiar with Tunguska, the crash from Tunguska [1908 explosion in Russia]?

Grandpa: Yes.

MS: That was a crash then?

Grandpa: Yes.

MS: Do you have other information on other crash sites, how many ... where they are?

Grandpa: It is not important. But I know of ... crash sites, yes. Tunguska was ... it was not a manned craft, it was a ... what you call a ... it was a probe [with a] propulsion system that is [unintelligible]. Humans cannot understand ... human beings cannot understand it. Hmm ... lots of energy, lots of energy needed to ... we use it ... humans cannot understand yet; getting there, but humans cannot understand.

MS: Are you referring to Element 115?

Grandpa: That is ... that is part, yes, but ... just a part. It is an insignificant part; it is part of the process, yes. That is not the only form of ... travel ... transportation. There are many. There are many types of ... transportation ...

RN: You mentioned we are overpopulated on Earth.

Grandpa: Yes.

RN: We have approximately six billion people. What would be an appropriate balance?

Grandpa: Hmm ... less than half that.

RN: Thank you.

H: Grandpa, I think a lot of us would like to see you as Stan and Lisa have. Is there a way to invite that kind of experience?

Grandpa: If needed. Possibly, in the future. Maybe collectively.

LR: Grandpa, when you take Stan out of our house, how do you get him out with the doors locked? How do you do that?

Grandpa: Matter is not what humans think. Matter is a vibration. Matter is ... mostly—Stan knows this, matter is mostly hollow. We change the vibration ... and what seems solid is not really solid.

LR: Do you take him through the walls?

Grandpa: Yes. Sometimes. Sometimes we do not.

LR: Like at Robert's cabin when you take him back in and lock the door behind him?

Grandpa: Yes. I do lock the door, yes—for a reason, though. There was your government … monitoring them … they have tried to harm Stan.

LR: Are they the ones who called the cabin when we first got there? Or was that you?

Grandpa: No.

LR: Can you tell us why he comes back with bloody noses all the time?

Grandpa: It is … part of the process. It is … an unfortunate part for Stan, but … he must … be examined … and that is why he comes back … that way. We try to make his memories … we try to make it so there is no memory of experiences so the pain is not there for him.

LR: You may think this is a silly question, but my cat Lacy—she was around a lot. Did you have contact with her?

Grandpa: Yes.

LR: Do you know where she is now?

Grandpa: I do not. I do not know.

LR: Do you have any questions you want to ask any of us?

Grandpa: No.

SD: Grandpa, you said that with my experience—that it was with a different group? My phenomenal experience I had last year. Can you tell me about that group at all?

Grandpa: It is not important.

SD: Can we relocate as a human race?

Grandpa: You are not ready. We will not allow any hostile races to continue to be hostile. Humans must grow up.

MS: Down at Daniels Park when I was waving my lights, did you take that as an aggressive movement?

Grandpa: I have heard of this, yes. I was not there, but others monitoring at first thought it was aggressive, yes.

MS: You don't *now*?

Grandpa: Hmm … we do not find you aggressive, no.

Attendee: Can you comment on man's evolution about expanding from a two-stranded DNA to a twelve-stranded DNA in the etheric realm?

Grandpa: Man has been designed ... we believe to hold something greater. We are ... *word* ... we are studying. Hmm ... we do not quite understand where or who they are ... as advanced from us as we are from you. And ... it seems that man is a conglomeration of different races ... and we do not quite understand it yet.

Attendee: So those beings that created us are no longer involved in our world?

Grandpa: No, they are, but we do not understand it.

Thomas: You do not know who they are?

Grandpa: We know of them, but they are advanced. As advanced ... to us as we are to you.

Thomas: Thank you.

H: You said the triangle on Lisa's head was for monitoring. Does that mean she has an implant, too?

Grandpa: Yes.

H: Do any of the kids?

Grandpa: Hmm ... no.

LR: Can you tell us what the significance of the writing on [Stan's stepson]'s wall is?

Grandpa: Validation.

LR: For who?

Grandpa: For Stan and for you.

LR: Because no one can understand what it says.

Grandpa: It is just validation ... it is a minor validation. [Stan's stepson] was chosen because he ... is with Stan. Everyone that has been taken has been taken because they are with Stan ... but they are not *of* Stan so there is no importance.

LR: And was [Stan's stepdaughter] taken when you came to our house and left a circle in the yard? Was she taken with Stan that time?

Grandpa: She was taken ... no, she was taken out in the field, and it was to see if there were any similarities, but there were none.

LR: In the churchyard?

Grandpa: That is a church ... church in the field ... with the light, with the light ...

LR: The church light.

Grandpa: The churchyard then.

AR: In the equation there's also a symbol. A circle within a circle, with an arrow coming out of the inner circle. What does that symbol represent?

Grandpa: That symbolizes forward movement ... forward propulsion. The circle represents stabilization of forward propulsion.

AR: When Thomas and I were in the San Luis Valley this summer, there were beeping noises coming out of nowhere. Was that you?

Grandpa: Yes, a form of monitoring ... there are things ... there are technologies we use that produce that type of ... sound. We do not hear like you guys ... .guys, guys ... Stan's word, "guys." Humans can pick up on sounds that we do not hear. We ... wish we could hear.

AR: A couple days later, we heard voices in the woods. Was that you, or do you know about that? Do you know of that incident?

Grandpa: No, I do not.

AR: And then, human energies: Do you picture them in the shape of a "merkaba" or any sort of energy in the shape of a merkaba around humans?

Grandpa: I am not familiar with that, no.

Leo: Energy field ... or ... aura around a person?

Grandpa: We perceive differently than humans. We do not think like humans. We ... we ... physically see different wavelengths. We do not communicate like humans. We do not ... we perceive ... through here (touching the center of forehead).

Leo: Mental ... perceptions?

Grandpa: Hmm ... yes.

AR: When you were monitoring Thomas and [me] in the San Luis Valley when we first heard the beeps, why were you monitoring us?

Grandpa: Hmm ... briefly we monitor to see how ... to ... monitor people involved with Stan. Sometimes because we are ... is to see ... what experiences they are experiencing at the time.

AR: Again, thank you for your answers.

H: Recently Stan saw a government truck pull up and install something on the light pole right near his home. Do you know what that is?

Grandpa: I do not know, I do not know. Stan is being monitored very, very, very closely by ... not necessarily hostile intentions but ... paranoia is a bad thing. (*Looking at Paul*) Paranoia is a bad thing.

AR: The letter that was spoken of before, with the ancient writing on it? What culture was that from?

Grandpa: The letter that Stan wrote?

AR: No. It was a letter that was given to Stan ... that had ancient writing that you said was for confirmation ... to help? Stan wrote it? An equation that Paul witnessed ...

PW: The bottom line of an equation was ...

H: Someone said it was Aramaic.

Grandpa: It was an ancient language, and it will be know soon what it is. It is for validation.

AR: It is an unknown culture?

Grandpa: No, it is not unknown; it is just ancient.

AR: Do you want Paul to know that paranoia is not good?

Grandpa: Paul should know this, yes.

Group: (*Laughter*)

PW: Thank you.

Grandpa: There are others that have questions that they want to answer.

Attendee: Grandpa, when we generate emotions, can those emotions be collected or utilized for anything?

Grandpa: There are some that can do that, yes. There are some ... some that are not of good intention that ... would feed off of negative emotions ... they are not a strong species ... they are ... a weak species ... and ... emotions can ... it is hard sometimes for us to deal with your emotions. Your emotions are too strong sometimes.

Attendee: How do they affect you?

Grandpa: Sadness and fear ... are painful for us.

Attendee: Is love exhilarating?

Grandpa: Love is a positive emotion. Love is a ... universal truth. Most intelligent civilizations try to love. They know this to be the correct direction.

Attendee: Thank you for sharing that. Are you aware—I'm fascinated as many are—of the fleets of ships or lights that appear in the sky with fair regularity in Mexico? Are you aware or do you have an impression that they are connected to you or—

Grandpa: Contacts have already been made. Contacts will continue to be made as man shifts ... as man's consciousness accepts that there [are] others out there. Some of what you see is not us. Some is ... there are ... creatures—not creatures, living things that are ... they are natural to what you call space ... they are ... not intelligent and reside in what you call space.

Attendee: You term them biological?

Grandpa: Biological, yes.

RM: So there are many different craft that we see in the skies?

Grandpa: There are *many*.

RM: Are they all from different species?

Grandpa: Some are from different species. Some are like you have here ... models of transportation. You have different forms in your models you ... so do other races. There are many.

RM: How many races are flying craft over our world?

Grandpa: Many.

Attendee: How many beings from your civilization are here on Earth? How many craft do you have?

Grandpa: It is not important.

AR: Are you familiar with [name withheld]?

Grandpa: Yes.

AR: Is he still receiving visitations?

Grandpa: I do not know.

AR: Did he ever?

Grandpa: Yes.

AR: Are there many human-appearing beings coming from other planets or systems?

Grandpa: There are beings that are ... resemble humans, yes. There are beings that ... do not.

RM: Are you familiar with the creature we call Reptilian?

Grandpa: Yes.

RM: Can they change shapes?

Grandpa: Hmm … maybe not necessarily change shapes … they … *word* … they *con* … they … can project … con people, con humans … Stan's word … *con* people into … seeing something different.

Leo: Different perception?

Grandpa: They are able … it started out as a defense mechanism. They can … project their image onto humans, and humans will see what they want them to see.

Attendee: I know everybody has asked you this question, but I was wondering, is the only reason I'm here because I am related to Heidi, or—

Grandpa: No.

Attendee: Is there a reason?

Grandpa: Yes. You will help … get the word out. Being interested is … positive and … you will get the word out …

Attendee: I don't know much about it …

Grandpa: You will learn. You are more than you know.

AR: Are there areas in Colorado and New Mexico you would recommend we stay away from … that are dangerous?

Grandpa: Colorado Springs.

Group: (*Laughter*)

Grandpa: Not because of us, but because of … um … men … (*slowly*) *military* …

AR: Air Force operations? Special investigations?

Grandpa: They are … the ones that are … bothering Stan and Lisa.

AR: Are there offices out of Shriver Air Force base?

Grandpa: There are many offices including many locations, but yes, that was one of them.

AR: Thank you.

MS: Stan has a biological daughter. Can you tell us if she's been abducted?

Grandpa: Yes. She has. [Stan's daughter]. She is of interest to us.

MS: Because of her mental abilities?

Grandpa: Yes. She is of Stan. She is … like Stan. She is naturally like Stan. Stan has a unique design. [Stan's daughter] is not … is not designed, but is *of* Stan.

MS: Do you know if there is a chance of better relations with [Stan's daughter]?

Grandpa: It is not up to us. We cannot interfere in that way. It is tragic that Stan has had to go through what he has, but we cannot interfere. When [Stan's daughter] gets older, Stan will have a connection with her that will be natural and very strong ... but she must get older.

Attendee: On our way up here today, we saw helicopters, and I was just wondering if it had anything to do with Stan?

Grandpa: I am not sure.

SD: Grandpa, I have a request. Is there any way that we can get more evidence or more information to expose this case to the public and make it more believable and more real? Can we get physical evidence?

Grandpa: There will be more physical evidence when needed. We cannot ... we can guide; we cannot get involved. There are rules. We cannot ... humans want sovereignty ... humans must find their own way. They must grow up.

LR: When Stan brought some "stuff" back in his hand, did you allow that ... [The "stuff" refers to a metal foil that I brought back from an abduction.]

Grandpa: No.

LR: ... or did he just take it?

Grandpa: Yes. Stan is smart. Stan is smart.

LR: Do you know of ghosts, as we call them ... ghosts or spirits?

Grandpa: Yes.

LR: So can you tell us if last November, if it was a ghost in our house, or if you were there with us for Thanksgiving?

Grandpa: Both. It seems that because Stan is different, he is bright. He is bright ... he is bright, and there are those that are attracted to him because he is bright.

LR: So are they able to come through because you come through? Do they use the same gateway?

Grandpa: Hmm ... we believe so. We are studying this phenomenon. We do not understand it very well, but ... it seems that we know humans are more than they appear. Humans are ... humans don't know what they are.

LR: What are we?

85

Grandpa: You are more. You are more than your physical side. You are more. You are a conglomeration, but you are more than that. We are studying this ... we are studying this.

RN: Since you are a researcher, studying humans, what has impressed you most about humans?

Grandpa: They are quick learners. They are ... able to ... *word* ... they are able to coexist with creatures not like themselves. They ... communicate ... it is part of their ability that has been ... subdued, switched off ... but they have the ability to coexist with animals and pets. *Word* ... most civilizations ... some do, but mostly they are not indigenous species. Also, humans have the ability to ... what they can imagine; they can create. They are quick learners.

MS: I just wanted to quickly find out if there's some significance to the Bentwaters incident [1980 UFO sighting in England]?

Grandpa: Yes. Yes, it was just the government. There is an ongoing contact with certain parties that want this stopped because the information that should be shared is not being shared. And ... things do not [unintelligible] he will be given a gift ... he will get to see.

Attendee: Can I ask your opinion on what happened September 11?

Grandpa: It was ... deceiving. It was ... lies ... it was for ... personal gain. It was not what ... it appears to be ... [unintelligible].

Leo: I have a question. If Stan and Lisa are willing, are you also willing to share more information openly?

Grandpa: If needed. If allowed, yes.

Leo: We do thank you for being here. We thank you for the information. We appreciate your doing this. Thank you.

End of regression session two.

Chapter 3

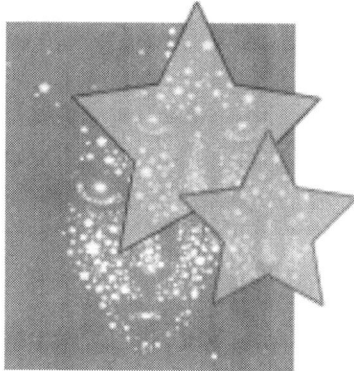

REGRESSION SESSION THREE

FEBRUARY 11, 2007

To protect people's privacy, there are times when we have to use pseudonyms. In this third regression session, one of the top scientific researchers on the topic of crop circles was present. We used the name "Susan Carson" to represent this person. Also present for this session were those of my inner circle, who also were encouraged to ask questions.

The regression session begins with Dr. Sprinkle's first question.

Dr. Leo Sprinkle (Leo): Is Grandpa here?
Stan Romanek (SR/Grandpa): Hello.
Leo: Hello. Welcome. Thank you very much for being here. And if you are willing, we'd like to have an opportunity to ask questions with

Susan Carson, who's here [and] would like to ask some questions. And if you're willing to provide information, we'd appreciate it very much.

Grandpa: Yes.

Leo: Thank you. And it's OK if Susan asks these questions and if you respond to them through Stan?

Leo: Yes.

Grandpa: All right, thank you. OK, Susan, whenever you want to, you can just go ahead.

Susan Carson [pseudonym] (SC): Apparently you know of me.

Grandpa: Yes.

SC: And I wonder, how?

Grandpa: You work with Stan; you have been chosen. We know of you before Stan.

SC: Why? Why? Why me?

Grandpa: You are ... curious. You are ... inquisitive. You are ... you have the potential to guide ... you do not ... jump to conclusions. You seek truth. That is why.

SC: Can you tell me anything about yourself?

Grandpa: If need be, a little.

SC: I think it would be very helpful to me. As you know, I certainly must know. I don't understand this, and I do want to understand it. So knowing *something* about you would probably be helpful.

Grandpa: You can ask your questions.

SC: Do you know ... are you aware of mankind's situation generally?

Grandpa: Yes.

SC: Is this because you have the ability to ... read the minds of ... perceive what's in the minds of mankind?

Grandpa: We can, but it is not necessary in most cases. We monitor. We can see that ... mankind is on a destructive path, mostly not because of their own doing—mostly because of leadership and ... because they are guided incorrectly. It is ... our purpose to enlighten. It is our purpose to ... show them the correct path. It is *your* purpose to ... when you come to terms with it ... and you believe that what is happening is real, you will guide also. But we leave that to you to decide if that is what you want to do. We will not force ourselves on anybody.

SC: One of the things that interfere with my ability to accept at face value what other people in this room have accepted at face value ... is that I know that the human personality, in order to communicate and share, has to trust. Trust is a critical issue.

Grandpa: Yes.

SC: Many of the people like Stan who have experiences with consciousnesses, beings of your sort, have these without their permission. The encounters occur without them being aware—without them being able *consciously* to say, "Yes, I'm willing." This worries me enormously. Can you talk about that?

Grandpa: It is a double-edged sword. Humans want to trust, but humans cannot be trusted. Humans experiment on animals and do not care ... guide. There are others that do not care what humans want. And ... they are not us.

SC: All right, you say that humans experiment on animals. Yes, some humans do. Other humans do not, and would not.

Grandpa: We are aware.

SC: Then ... if you know that some of us are not of that sort, and some of us in this room are *definitely* not of that sort, why can't you appear completely openly to people like us?

Grandpa: Humans are not ready. Humans are still primitive. Humans are ... word ... hostile ... although—

SC: Frightened?

Grandpa: Hmm ... frightened, yes, but their reactions to fear can be hostile. We are trying to ... *word* ... we are trying to ... (*slowly*) *slowly*, slowly ... prepare humans for the shift—contact—for contact. Once humans ... believe they are not the only ones, the shift will happen. You are part of this. You are ... to help in this. It is up to you. We are waiting ... we are waiting for you to accept, for this to move forward.

SC: Thank you. I have had—as you know, I suspect—many unusual experiences.

Grandpa: Yes.

SC: One of the things I observe is that often these very unusual experiences occur very suddenly, very abruptly, and quite dramatically in the physical realm. This of course, is startling to humans not prepared for it. Why is it not more gentle? Why are these ... awakenings, these demonstrations ... why is no warning given so that one can prepare oneself?

Grandpa: It is a matter of timing. There are some that are ... involved that are not us. We are involved, but there are others that are beyond us that are showing themselves to you because they know that your involvement in this situation is important. And ... *word* ... it is a matter of timing. When things are right ... they have to ... Stan's words ... go with (*slowly*) *the flow* ... go with the flow.

SC: OK. Are there guidelines that you can suggest ... which I and others could use to recognize you and your kind of individual as separate from others? How can we tell the difference?

Grandpa: Hmm ... it is ... comes natural. Humans, humans have instinct—it is not instinct; it is beyond instinct. You can tell if a person is of evil intention, and you can tell if a person is not of evil intention. You have that. Humans ... in this room especially, have that ability to decipher what is good and what is not. You will know when it is us and if our intentions are of pure thought, and of ... *word* ... loving intentions.

SC: I feel great love around Stan.

Grandpa: Yes.

SC: In the people around Stan.

Grandpa: Yes.

SC: It actually feels to me very much like the love that my friends and I experienced in the bluegrass music scene. It's very similar, isn't it?

Grandpa: I am not sure what bluegrass, bluegrass ... I ... music. Stan ... things have been put into place to support Stan and his family. What Stan is going through is very hard. It would be very, very hard for any humans. Stan's ... purpose is to lead, to guide. He is a beacon. He will be a beacon. He will bring them to the truth. He must get help. He must be supported. There are those in ... *word* ... there are those in your race ... there are those of you that do not want this for ... greed and for ... *word* ... selfish reasons. There are also those ... up there as there are down here. It is no different. What humans are ... stuck ... they are ... they cannot see ahead. They are stuck in this reality, and they are guided incorrectly. We are trying to change that, but ... we have rules; we have laws, just like humans do. They cannot conceive of what we conceive. But we cannot directly get involved. It is not allowed; it is forbidden. But we can guide. That's what our purpose is, to guide.

SC: I have a couple of hard questions here. Humans, as far as I can tell, did not create themselves. Humans ... we are dealing with ... the cards we're dealt. How we're born ... we have to deal with the cards that are dealt. Where did we come from? Where do people come from? And who did it?

Grandpa: We are ... studying this. We are not ... Stan's words ... (slowly) exactly ... exactly sure. We are studying this. We ... are certain that humans are a (slowly) conglomerate ... conglomerate ... of different species for purposes yet unknown, but ... humans are more. Humans are physical and spiritual. They are a combination. They are a water vessel. They are ... word ... they are something more. Humans do not understand who they really are.

SC: Who ... do you have any idea where we come from? Are we of the Almighty?

Grandpa: There is truth of this. There is a ... as we are advanced to you, they are advanced to us, and maybe even more so. They are ... greater.

SC: Are you of the Almighty?

Grandpa: We are part of the galactic ... we are part of the consciousness. We are physical, but we also can perceive of ... the true nature and the true ... the truth of ... the word ... cosmos. We are a part. We are ... we are all a part. Humans, us, everything is a part of ...

SC: So we together are of the same source. Is that what you're saying?

Grandpa: That is correct. There seems to be a oneness that has fragmented itself. We do not understand; it is beyond us. We believe that it is to ... experience. Humans are here to experience. We are here to experience ... but to what degree, we do not understand. It is about the experience, but ... love is ... the way of it. Love is ... the way it should be. Stan knows this.

SC: Is your interest in guiding humans for the benefit of all, or more specifically, is it for the benefit of your particular race or some other race?

Grandpa: For the benefit of all, this what we do. We ... have chosen to do this. Humans are a distant race. Humans are more. Humans are a water vessel, but more. Humans ... humans do not know who they are. They are greater than they even... think, think. They are greater than they know.

SC: I was just thinking that I was going to ask you what are perhaps the finest aspects of human beings?

Grandpa: Humans have the ability to ... cohabitate—coexist with other creatures. Others have this, but humans are best at it. Humans can understand ... how others ... feel ... feel ... coexist. They can coexist. Humans are more than they appear. They have a divinity that others ... they have a divinity that others do not.

SC: Do you have that divinity also?

Grandpa: Hmm ... humans are different. We are studying humans. We are guiding humans. We are waiting for the shift. The shift must happen or humans will destroy themselves. Um ... we are different. Humans are different to us.

SC: So you do *not* share in this divinity?

Grandpa: We do, but in a different way. We are divine; everything that is created is divine. Some have lost their way. We ... do not want humans to lose their way.

SC: On a different note, if it's all right—can you discuss crop circles?

Grandpa: Yes.

SC: Are they wonderful invitations to learn?

Grandpa: They are ... some are not us; some are you. Some are humans'. Some are ... it is meant to ... guide ...

SC: Guide?

Grandpa: It is meant to ... subconscious ... subconscious ... it is a ... there are others—not just us—that do this. It is to ... get them interested. It is to help humans slowly ... to move forward, to understand that they are not alone; they are not the only ones.

SC: We have some evidence that crop circles did exist as long ago as five hundred years ago.

Grandpa: Older.

SC: Older? Was the occurrence of those crop circles five hundred and more years ago the same as the purpose of the appearance of crop circles now?

Grandpa: No. Some are ... circles ... circles ... circles ... some are natural; some are not. Some are man-made. It all serves the same purpose now. It is to enlighten. It is to spark interest. It is for that purpose only that they are created now. Before, it was ... it was a beginning, but the purpose is stronger now.

SC: Is there greater urgency now, as many people think?

Grandpa: Yes. Greater. Greater urgency.

SC: Why is it more urgent?

Grandpa: Man is at the brink of destruction. Man must —as Leo puts it—"grow up." Stan knows this. You know this.

SC: Given the fact that modern culture relies upon the authority of science to accept almost anything nowadays, is there a way to involve more scientists so that more people can be included in this awakening, in this enlightenment?

Grandpa: That is what we have tried to do. That is what Stan is doing. We have given Stan information that is ... not readily available, to pique the interest of what you call scientists. Humans have been guided incorrectly. Humans are ... being guided by evil intentions and greed and ... this must stop. Humans are ... *word* ... standstill, *(slowly) standstill* ... at a standstill in their evolutionary development, physically and mentally because of this greed and this ... this ... "leadership" ... a leadership that has now become an *(slowly)* *epidemic* throughout the world unless it's taken care of ... and if they are not—soon—humans will destroy themselves.

SC: I just had a very interesting idea. Would you, are you able to draw me a crop circle that is a genuine crop circle—not a man-made and not made by people of bad intent or creatures of bad intent? One that is done to cause people to open their eyes. Could you draw me such a one?

Grandpa: Hmm ... I can. I have ... I do not make crop circles, but ...

SC: Would you know of one that was this way?

Grandpa: Yes.

SC: Would you draw it?

Grandpa: Yes.

SC: Oh, I would love that!

Leo: Here's a marker pen and a pad. (*I am handed a drawing pad and marker. With eyes closed, Grandpa begins to draw.*)

Grandpa: I will give you one in the future.

SC: Fine ... (*I finish drawing, and the sketch is shown on camera.*)

Leo: Thank you very much.

SC: Oh, how pretty! Oh, pretty! That's quite lovely! And it's quite unique.

Grandpa: That way, you will know it is from us.

SC: Uh-huh ... definitely our boys.

Grandpa: And we will try to ... we will wait for your visit with Michael [pseudonym].

SC: Ah ... tell me about Michael—what you feel is important to know about Michael?

Grandpa: He is of the seven. Stan is of the seven. Michael is a piece. Michael is ...

SC: He was recently very, very frightened. Do you know about this event?

Grandpa: Yes, we know why.

SC: Why?

Grandpa: Because there are those that do not want this ... up there as there are down here. Michael is ... experiencing their hostility. But Michael will be fine. Michael is stronger than he knows. Michael is ... they are not working with Michael. Higher, higher ... are working with Michael. We have visited Michael, but he is a conduit. He is a very colorful conduit. He is of the seven. You know of this. You will guide him. You will lead them. They are waiting for you to realize who you are ... before we ... before we move forward with this.

SC: Are there multiple consciousnesses approaching Michael?

Grandpa: Yes.

SC: Some of them are not of the highest interest ... I believe. I think some around him are ... perhaps intrusive?

Grandpa: Yes. Some are just lost. Michael is a light like Stan. Michael is a stronger light than Stan, but a light nevertheless. Michael attracts things he does not understand because he has an ability to ... *word* ...

he has an ability to … Michael … has been chosen. He has been altered. Stan has been chosen; he has been altered. Everybody involved with Stan and Michael has been chosen, and they have been altered. But there are seven. The seven must get together. The seven must get together to experience each other. Um … the power will increase … exponentially.

SC: Michael tells me that there are seven also. He has told me that for some time. He does want to meet the others, but his situation is very narrow. It's the control of his father.

Grandpa: We are aware of this. It is not … we are not in charge of Michael. Others are [unintelligible] also. It will be OK. Michael will … be fine. Um … you are speaking of Michael's parents.

SC: Yeah.

Grandpa: Michael's parents … are also there for a reason. There are those that will try to … *word* … there are those that will try to … confuse and … misdirect Michael. Michael's parents instinctively know this and are very protective. And … can sometimes be overprotective, but … Michael is beyond this. He is, therefore he needs them. They are there for a reason. When the time is right, things will be what they are supposed to be.

SC: I tried to help Michael recently. And I told him he and I together are trying to teach him how to do something in particular to help his situation. I wonder if you know what that thing is, that we are trying to do, and if I was correct? Was it a good thing for me to try to encourage him to do?

Grandpa: Hmm … I am not sure, but he will learn to protect himself, as … if needed. He has protection around him. There is a battle between what you call "good" and "evil," and Michael knows this instinctively. He is not without protection; if that is what you are talking about, he is … very strong. Stan also has guidance and protection. You also have guidance and protection. Our intention is not to overwhelm. People involved with Michael's intention is not to overwhelm. Michael … will handle this as he is able—no more, no less.

SC: Do you know about a man in Italy, a psychic, a medium whom I met on my trip recently?

Grandpa: He is … um … he is …

SC: Near the Mediterranean.

Grandpa: You know of the seven. I believe he is the one you are talking about.

SC: He was suggesting that ... he had many questions about Michael. [Name withheld]. His name is [name withheld]. And ... I was very confused by the questions he raised about Michael, as if Michael was not behaving properly. I'm very confused about this. I don't know who to listen to.

Grandpa: That would be a ... Michael is ... instinctively ... Michael knows what is correct. There are ... bad and there are good. Bad will try to disinform; bad will try to confuse. Michael's intention is to ... heal, to help. Stan's intention is to help and to heal. The human race is ... at the brink of destruction. The human race must be helped. We cannot get involved directly, but we can guide. Our help is to use ... us and others ... our help is to use ... humans that are willing and that are ... different ... to help man do this. Michael is one of them, the seven. There are seven. You know of the seven; you know who they are. We are waiting for you to decide what your place is in this. You are ... unsure. And we are going to be patient, but we cannot be patient too long because ... man is coming to the shift. If the shift does not happen, man will perish.

SC: What would be lost in the universe if mankind did perish?

Grandpa: Man belongs; we all belong. Man is divinity—part of divinity. We are not part of that divinity. Man is ... purposely to understand. We are not sure of the—

SC: I think I see.

Grandpa: ... pre-cussions [repercussions]. Man is a piece that is needed. Man is a young race. Man is a violent race. Because they are young, they are primitive, but they have potential even we do not understand as of yet.

SC: Is there something, anything, that gives you ... is pleasurable or rewarding to *you* that mankind is capable of?

Grandpa: Stan ... can be fun ... funny.

SC: Funny? You enjoy humor?

Grandpa: We play with Stan. Humor—yes, humor is everywhere, but we do not think like humans. We are not emotional like humans. ... we do not communicate like humans.

SC: I was going to ask you that question.

Grandpa: We are communicating through Stan ... human communication is slow.

SC: You communicate in concepts. Yes?

Grandpa: And more.

SC: Whole scenarios at once.

Grandpa: Exactly. We are more advanced because we can communicate more advanced. We ... speaking ... Stan knows this; he talks about this ... speaking is slow. And we can communicate ... faster.

SC: Can humans learn to do this the way you do?

Grandpa: Humans do have ... humans have been turned off. We do not understand who or why this has happened. Humans have been turned off. *Word* ... they have the ability ... there are hints of this ... humans do not know who they are. We do not understand ... humans have instinct; it is more than instinct. Leo Sprinkle knows of this. You also have instinct that is greater than just instinct, Stan also. There are ... confusing ... we are learning.

SC: Ah, you also learn.

Grandpa: Yes. Everything learns. There are things greater than us ... Stan has talked of this. There are ... the more intelligent ... a more, greater ... intelligence. The more intelligent the rest [unintelligible]. Humans may learn. Humans will not stop. There is a universal law that's keeping the universal law.

SC: In your efforts to communicate or guide human beings, I would like to remind you—reiterate—that humans, when they are frightened, *cannot* help how they respond.

Grandpa: They must learn.

SC: They must learn?

Grandpa: They must grow up. Part of their ... insecurity causes their ... their ... reactions.

SC: But it's *instinct* to respond ...

Grandpa: Instinct, yes. But ... humans are primitive, and they feel primitive response.

SC: But we can't help that! That's how we are!

Grandpa: I understand, but they must grow up.

SC: One has to learn to get past that?

Grandpa: Humans have … humans have more control than they think. Humans tend to be … they … they …

SC: Yes? Yes? Yes?

Grandpa: But they must understand that their survival depends on them to be more aggressive in this way …

SC: A-ha …

Grandpa: They must learn to subdue their fear and their anger.

SC: Fear.

Grandpa: It is OK to be … cautious. You are very cautious. It is not good to be paranoid. Some of Stan's friends are paranoid. It is not healthy. Paranoia … (slowly) paranoia breeds … Stan's words … breeds … breeds contempt.

SC: Oh, yes. Stan and I need to understand … when you appeared—I think it is you I'm speaking with—when you appeared outside the window and startled Stan so badly …

Grandpa: (Laughs)

SC: That was funny?

Grandpa: (Laughing and grinning)

SC: Why … like that? Why? Because he was so frightened?

Grandpa: Stan is … more controlled than you realize. Stan was frightened, yes. But … he was not terrified. It was an experiment … to see how ready Stan was … for this.

SC: A-ha. Is it possible for you to … is it difficult for you to materialize physically?

Grandpa: Hmm … you do not understand. It is beyond humans. Out technology is beyond humans. There are things you have not [unintelligible]. You are not ready, and it is not allowed. It is allowed when you are … (slowly) experimenting. But it is not to impact human consciousness. Humans must … we have given Stan information to give to your scientists. That will help guide them. It is also confirmation that what Stan is going through is real so others may know that this is real … and may come to the understanding that they are not alone. And when they realize they are not alone the shift will happen.

SC: You are aware of Dr. [name withheld] and Dr. [name withheld]?

Grandpa: Yes.

SC: Is there a difference between those two men that we should know?

Grandpa: [Name withheld] is of ... honorable intentions. [Name withheld], we have found, is not of honorable intentions. I would not suggest staying with [name withheld] ...

SC: Thank you.

Grandpa: Although ... this is for everyone, do not give [name withheld] information.

SC: Do not ... do not *what?*

Grandpa: Give [name withheld] information.

SC: I know what I felt but ... I am not always *sure* what I feel is accurate.

Grandpa: There are those that are here to deceive. There are those that have been hired to deceive. There are those that are here to ... *word* ... of their own intentions to steal, to ... *word* ... to say it is ... *word* ... to say it is of their own doing, but it is someone else's ... they do not revere.

SC: Is it maybe time to take a break? Is it tiring at this point?

Grandpa: Not for Stan, but if you wish, we can take a break.

SC: *(To Dr. Sprinkle)* Is it all right with you? Are you OK?

Leo: Sure. M-hmm.

SC: Everybody's OK—I don't want to ...

Leo: Stan's OK.

SC: *(Long pause)* I'm thinking ... I ... um ... have had a sense for quite a long time that ...

Grandpa: Yes.

SC: Ha-ha ... that ... I have a job to do.

Grandpa: Yes.

SC: And ... very specifically ... that I was to try to reach the people who require the authority of science ... to be given permission to ask bigger questions, and that's what I've done, what I've done.

Grandpa: Yes.

SC: Because I'm not a scientist myself, I can't always judge which of these people are actually useful—helpful to this cause—to the purpose.

Grandpa: You must be guided by instincts. Your instinct is correct most of the time. It is not a ... *word* ... mistakes are part of the experience. If you do not find somebody right, you move on. Stan knows this. He will be guided correctly. You must decide what your role in this is. We are waiting for you to decide. You know instinctively inside that you are ... one of the seven. We are waiting for you to decide if that as well. We do not want to push you.

SC: It's very difficult for me to imagine. I don't know what to do. I don't know what to do.

Grandpa: Yes, you do. You are doing it. You know exactly what to do, but you do not trust. You do not trust your instincts.

SC: Do you have a magic mirror? "Mirror, mirror on the wall ... what am I to do this fall?"

Leo: (*Laughter*)

Grandpa: You are doing what you are supposed to. You are supporting Stan. You are more than you know. Stan is more than he knows. Michael knows who he is. Stan will be a leader. Michael is who he is. You are ...

SC: Do you know about the man in Poland?

Grandpa: I know of him, yes.

SC: And ... is he as I *feel* that he is?

Grandpa: Everybody you feel is are [unintelligible]. You are correct [unintelligible]. You know of the seven; you know of six, but you do not understand yourself.

SC: Why does the man in Poland sometimes respond to me a lot and other times not respond to me at all?

Grandpa: He is confused just like you are.

SC: Ah.

Grandpa: But we are not [unintelligible] to ... we are here to guide; we will not ... push. You have to accept us yourself like Stan has, like Michael has ... and like others have. You know of these people. Your goal is to ... your *involvement* is to ... connect everybody, to get them together. It is important. It is imperative to get them together. When the timing's right, even Michael will understand this and will be involved. You are compelled to do what you do. Michael is compelled to do what he does. Stan is compelled to do what he does. The others are compelled also.

SC: Are we compelled because of our ethical and moral and spiritual and intellectual development?

Grandpa: That is part of it; there is more. You are ... there are different keys to open a ... *word* ... large lock ... Stan's words.

SC: Large lock. (*Laughter*)

Grandpa: *Word* ... there are ... each person involved has their role. There are seven; you are one of the seven.

SC: These seven together ... once together will be ... more powerful? More effective?

Grandpa: They will understand what to do. They must get together to ... experience ... to know ... *word* ... I cannot explain. You will see. It is for the betterment of man. It is ... they will help guide.

SC: (*To Dr. Sprinkle*) I need to think a little bit about what he has told me. Would it be all right if we took a break?

Leo: Sure.

SC: Thank you.

Leo: Thank you very much, and we'll take a little break.

After the break, Dr. Sprinkle induces me once again into hypnosis.

Leo: And Susan has more questions for you, if you are willing to respond to the questions?

Grandpa: Yes.

Leo: Thank you.

SC: Earlier we talked about how we are both learning—humans and yourselves.

Grandpa: Yes.

SC: What do we call you? What is your—we're humans; what are you?

Grandpa: You can call me "Grandpa."

SC: But of what species or race or type or ... variety of consciousness are you?

Grandpa: It is not important.

SC: I would like a name to call you other than "Grandpa."

Grandpa: It is not important.

SC: Even if it's important to me?

Grandpa: It is not necessary for ... it is not necessary for this process. You can call me "Grandpa."

SC: *Stubborn.*

Grandpa: There are many species. We are one of the many.

SC: But you are biological?

Grandpa: Yes.

SC: Are you carbon-based?

Grandpa: Um ... yes. And no.

SC: Do you ... experience sadness?

Grandpa: We experience grief, not in the extent humans do. We experience emotions, not in the extent humans do. We are ... we communicate differently ... emotions ... primitive emotions ... we are beyond primitive emotions. We cannot have primitive emotions in order to communicate ... the way we do.

SC: All right. You must recognize, though, that humans are *filled* with emotions.

Grandpa: Yes.

SC: This is simply how we are. And that we can't, without a great deal of effort, divorce ourselves from those emotions.

Grandpa: Humans are more than they think. Humans have the ability to control ... but humans have become lazy. Humans ... are by nature—*not* by nature—they have been taught to become aggressive. Humans are fearful when ... it is not necessary. Humans do have more control than they think.

SC: Why do you say that humans are lazy?

Grandpa: Because it is so.

Group: (*Laughter from other attendees*)

SC: And what do you mean by lazy?

Grandpa: Um ... *word* ... lazy ... they do not ... humans do not put a lot of effort into controlling their emotions. They are ...

Group: (*More laughter*)

Grandpa: ... all over the place.

Group: (*Much laughter*)

Grandpa: Their thoughts are all over the place. They are not ... *word* ... and it might not be only because of laziness—it might be ...

102

because they were not ... they are not taught correctly. They are not ... they do not know how to ... process correctly.

SC: Do you understand the word "inspire"? Inspiration?

Grandpa: Yes.

SC: I think that humans are motivated to focus, to discipline themselves ... *when* they are inspired.

Grandpa: We find this is true.

SC: Perhaps to facilitate ...

Grandpa: Many times we have ... given humans ... answers to questions ... that they think is ... *word* ... *word* ... that helps to inspire. They do not realize it is from us, but they think that ... *word* ... that they think it is of themselves ... and ... it is inspirational. We have done this many times. Humans ... have also done things ... on their own, that is ... they do not give themselves credit for.

SC: In your interest to guide human development and growth, I feel, that to the extent that you can inspire humans ... you would help ... you would help facilitate this growth; that ... inspiring them ... fear *dissipates* under inspiration. When humans are inspired, they tend to *not* be fearful.

Grandpa: There are certain things that we are not allowed to do. We cannot directly get involved. We can lead in such a way that is not ... does not ... allow direct involvement. Humans ... must reach ... enlightenment on their own ... but they can be guided. There are those that do not see it as we do and will try to force the issue in ... a negative way. We are not like this.

SC: So you're in opposition to ... or in battle with, to some extent with other forces—other entities—other consciousnesses?

Grandpa: Yes. There's a battle, yes. What you call "good" and "evil" ... it is more than you know. It is ... *beyond* good and evil. It is ... consciousness itself. It is ... *word* ... it is a ... oneness in war with itself. Humans do not understand.

SC: I have a faint grasp of it.

Grandpa: Evil is oneness in war with itself.

SC: Can you tell me, at the end of Stan's last equation [that] Stan produced; there were symbols at the bottom of the page.

Grandpa: Yes.

SC: And one of our group discovered that they represented an ancient language.

Grandpa: Yes.

SC: Why were the symbols chosen from that language rather than, for instance, English?

Grandpa: Hmm ... the symbols were chosen because Stan does not know this. It is for confirmation of Stan's experiences.

SC: I see. Was the fact that Jesus apparently spoke that language ... did that have anything to do with the choice also?

Grandpa: A small part, but, yes. Humans ... this information will get to more people, and when it does, there will be ... humans that will ... recognize that aspect of this information that was given to Stan and ... it will pique their curiosity. It will ... excite them in ways that ... we do not understand yet.

SC: Thank you. Are you aware of Robert?

Grandpa: Yes.

SC: Can you tell me a little bit about Robert ... why he is here, why he's part of the group?

Grandpa: He is ... here because of Stan. He is been chosen like everybody involved has been chosen. Robert is ... to get information out. Robert is a tool to ... get the information out to reach the masses. It will happen.

SC: You used the word "tool."

Grandpa: Yes. We are all tools.

SC: And from that meaning—and I understand, not in a totally mechanistic sense—are we *just* tools, or are we more than—

Grandpa: We are more. We are all part of the divine ... in different ways. We are ... all ... here to learn. We are all ... here to experience.

SC: Robert has some very specific questions about Stan and his encounters, which he would like to ask. Is that all right?

Grandpa: Yes.

Robert Morgan (RM): Hello, Grandpa.

Grandpa: Hello.

RM: I have some questions that are about Stan. The first questions is, Stan's body seems to affect electrical appliances.

Grandpa: Yes.

RM: Why is that?

Grandpa: He is different. He has a higher ... energy output. At first it was ... because of ... *word* ... you call ... track ...

SC: Implant?

Grandpa: Implant, yes. Thank you. But Stan is different; Stan has been designed to ... receive and to ... project energy for communication purposes. And his energy level is high, and it disrupts electrically ... sometimes magnetic fields.

RM: I put a camera on the side of Stan's house, and we got an event that we call the "siding sighting."

Grandpa: Yes.

RM: Are you aware of that?

Grandpa: Yes.

RM: In that piece of video ... we see these like little bubble things come in before the camera. Can you tell me what those bubble-looking objects are?

Grandpa: It is not important.

RM: There was an energy that was used to ... discolor the side of the house ... and take out the camera. Can you tell me what type of energy did that?

Grandpa: Hmm ... it is not allowed.

RM: Do you transport Stan from your ship to the ground with some kind of beam of light?

Grandpa: Yes.

RM: That's what makes the circles in the grass?

Grandpa: It is ... not light. Humans would not understand. It is ... it is energy. It is ... *word* ... humans do not yet understand.

RM: Are the light balls and light anomaly photographs—that Susan gets and Michael gets and many other people get—also of this other form of light?

Grandpa: Some are natural ... some are ... more. Some are energy ... some are purposeful; some are not. It is not the same as the energy we use. The energy we use to ... transport Stan ... is different. It is a ... not light; it is a ... matter, which you humans do not yet know about.

105

RM: Awhile back, Stan was riding his bike to work ... and some people pulled over and beat him up. Are you aware of that?

Grandpa: Yes.

RM: Do you know who those people were that attacked him?

Grandpa: We know of them, yes.

RM: Were they humans?

Grandpa: Yes. We are very disappointed.

RM: Are they our government?

Grandpa: Hmm ... not as you know of. *Word* ... they are ... part of the control. They are part of the ... they are ... against Stan. They are against what Stan stands for and what the seven stand for and what truth stands for. They are controlling for their own gain. They are ... of evil intentions. They want to control because they are afraid they will lose control.

SC: How did they even know about Stan?

Grandpa: They monitor Stan and ... they are involved very closely. They ... monitor constantly ... day and night. We are careful not to ... not to ... *word* ... *word* ... not to ... Stan's word ... not to appear ... when they're around. So they do not interfere.

RM: In Nebraska Stan was approached by a lady who drove up in a car, and she said, "If they want to get it right, you need to re-evaluate the Rosen Bridge" [Einstein-Rosen Bridge: geometrical property of a wormhole]. Are you familiar with that?

Grandpa: Hmm ... from Stan I am familiar, but ... it is all ...

RM: Do you know who that lady was?

Grandpa: No, I do not. We believe she is ... not *intentionally*, but still a part of the ... people who are against Stan.

SC: And these are human people? Regular, ordinary human people ... of our race?

Grandpa: Yes.

RM: Sometimes when Stan comes back, he's got wounds, and when we put what we call "black light" up to it, it glows. We call it "fluorescing." Can you tell us what happens?

Grandpa: It is not important.

SC: Is there a way that the scientists who are involved, is there a way that they could learn about this fluorescence? Is there a test that we can do?

Grandpa: It is not important. We cannot get involved directly. We try to minimize direct contact as much as we can. There is information we cannot give out.

SC: Is the fluorescence a result of the contact?

Grandpa: Yes.

SC: Is it a substance that is on the bodies of yourselves?

Grandpa: No. When the shift happens, all will be revealed.

RM: Years ago, Stan was at a hamburger restaurant, and a man walked up to him and said something to him. Are you familiar with this?

Grandpa: Yes.

Leo: Do you know who that man was?

Grandpa: Yes.

Leo: Can you tell me?

Grandpa: It was not me, but it was one like me. It was us ... warning Stan. We were ... not to scare him, but to prepare him.

RM: So you can make yourselves look like us?

Grandpa: Yes, if need be. It takes effort, but yes. It is a ... it is not important.

RM: Stan was also meeting some people at a coffee shop—

Grandpa: (*Holding up one hand to interrupt.*) It is a projection. It is a projection.

RM: Stan was meeting some people from the media at a coffee shop to talk about shooting his story.

Grandpa: Yes.

RM: And there was a man in that coffee shop with a computer ... what we call a laptop. Do you know who *that* man was?

Grandpa: We do not. We believe he was ... Stan's word ... (*slowly*) *surveillance.* Not us, but you.

RM: Now people keep trying to disrupt or attack Stan's computer.

Grandpa: Yes.

RM: Do you know who *that* is?

Grandpa: Yes, it is the evil. It is the ... controlling ... it is you, not us.

RM: Also, the people that did that story in May of 2002—the story was going to be on television. Right when the news came on, when that story was going to come on, the power in Stan's neighborhood went out.

Grandpa: We are familiar with this, yes. It was not us.

RM: I've had some things that happened at my house that were kind of strange. Is that your group's doing?

Grandpa: Some ... not many. Mostly ... yours—you. They were humans, mostly human. There are a couple of things that were us ... but ... mostly humans.

RM: I had a key that was hidden in a safety deposit box. Can you tell me where I hid that key?

Grandpa: Hmm ... no. I do not know.

RM: I also had some things missing from a closet. Can you tell me what was missing?

Grandpa: No. It is not important.

RM: Do you monitor me?

Grandpa: Sometimes. We monitor Stan ... in ways humans do not understand.

RM: Now you've been giving Stan dates by the alignment of the planets ...

Grandpa: Yes.

RM: ... several times. It that the date when mankind destroys itself?

Grandpa: Hmm ... it is not important.

RM: What is so important about the date that—

Grandpa: It is ... an event, a very important event. We cannot disclose this event. We cannot directly get involved. If we disclosed the event, we would ... directly get involved, and it is not allowed. We are trying to guide without getting involved.

RM: Thank you, Grandpa.

SC: This is a question from Robert and me, and others. How do we proceed? What do we do specifically? Are there things we can do— to facilitate, to accelerate the growth of people who are genuinely trying to communicate?

Grandpa: People involved must stay on the path they are on now. It will happen naturally. You are doing the right thing; Stan is doing the right thing. People involved in this are doing the right thing. It will happen naturally. We are ... interested in ... getting information to ... who will accept. But we cannot force people to accept. It is a subtle ... *word* ... it is a subtle ...

SC: Process.

Grandpa: Thank you. Process. It is a subtle process. And it must go according to plan. It must not be rushed, but it must be rushed at the same time. We need to not overwhelm people. It is hard for people to accept, but they must accept. And ... in order for this to happen, it must be done ... subtly. There is a ... human effort to ... get them prepared.

SC: May I ask you ... do you know ... do you have an opinion ... whether I am prepared—Susan is prepared?

Grandpa: We believe it is so. You are ... stressed, but not skeptical. You are not paranoid, but skeptical and concerned if this is of evil intentions or if this is of good intentions. You must not [unintelligible], but it is of good intentions. We are all connected; everything is connected. And if one thing goes away, it affects other things. If we go away, we would affect you. If you go away, you would affect us. If this mouse thing goes away, it affects ... called the ... Stan's ... Stan's word ... butterfly ... butterfly.

RM: Butterfly Effect.

Grandpa: Yes, thank you. Butterfly Effect. And it is so in this situation especially.

SC: I feel the honesty of that. I understand that. I feel that truth. I'm a little stymied.

Grandpa: You have concerns. You have worries. It is understandable. You must not worry. You must ... worry is ... subtle. If we really are ... you do not know who you really are. You have powers you do not yet understand. You will guide. You will help put together. You are the cohesive in the seven. You are there for a reason, like Stan is there for a reason, like Michael is there for a reason, like the others you know about are there for a reason. You have been given information. You feel that ... you feel that you will find the seven; you know who they are because you are designed that way.

SC: OK ... I think I understand.

Grandpa: You ... now know the ... reason why you are ... in the seven. Stan knows why he is in the seven. Michael knows why he is [in] the seven. It is your ... duty to ... explain to the others why they are in the seven and get them together. This is imperative. This is imperative for many reasons, but the biggest reason is for the survival of man.

SC: And you will help with this, yes?

Grandpa: Yes. We cannot get involved directly, but we will guide, yes. You have been guided; everybody here has been guided. It has been hard for us because of ... fear and distrust. We are sorry that the human race is so distrustful. But ... it still remains that they must grow up and learn quickly ... as they can because ... they are on the brink of destruction. It is evident in your weather patterns. It is evident in ... the overpopulation of the planet. It is evident in ... the pollution. You are killing yourselves and ... you are the only ones that can stop it. We cannot get involved. It would be tragic for the humans to die.

SC: May I take a totally different ... a *totally* different series of questions? It just came into my head.

Grandpa: Yes.

SC: The cattle mutilations.

Grandpa: Yes.

SC: Are they conducted—

Grandpa: You know. You know what they are. Yes, they are conducted. Some are us. Some are ... you. It is all the same ... we are experimenting for ... mostly the reasons you *know* of.

SC: We had some ... a lot of evidence, actually ... that indicates it's conceivable that the blood of these animals—the cow—is taken to be used by some groups ... in the raising of hybrid children. We know that people remove the heme molecule and produce bovine serum, which is then used in the development of fetuses outside of uterus here. We also know about Mad Cow Disease now ...

Grandpa: Yes.

SC: ... and wonder if the excisions are taken to test for this disease?

Grandpa: Yes. Some yes; some no. Some just to ... monitor environmental ... but ... we are not the only ones. There are human factors that also do this that look like what's ... relations ... we are monitoring ... there are reasons we are monitoring beyond ... reproduction. We ... are trying to ... (*slowly*) experiment. We are... learning as everyone must learn. We are gathering information about humans. Humans are an enigma; humans are different, but they are the same. They are a conglomeration of many species—one of this planet and many of other planets ... of off-world ... off-world.

SC: Why do you leave the bodies? Why do you not take the bodies and simply dump them in the ocean?

Grandpa: Why?

SC: I mean, is this to frighten ... you know?

Grandpa: It is ... efficiency—it's ... most of what you see, a small percentage is us, a larger percentage is human. Sometimes I believe it is to misguide ... to make fearful, and sometimes the intentions is just ... that ... efficiency ... *word* ... we do not have ... it was easier for us to just ... do what we come to do, to ... monitor and to leave. There is no reason to take with us when we are finished.

SC: I should tell you then, there *is* a reason because ... death is something which most people, most humans are afraid of. And dead animals cut up are ... shocking to humans.

Grandpa: Yes, we understand ... it is necessary in some cases. We do not intentionally ... try to kill or destroy. Humans kill daily. *Daily.* Humans are barbaric. We try to ... humans are confusing!

SC: As are ... creatures like you.

Grandpa: We are of pure intent. Humans do not know of what intent they are. Most of the times humans are ... confused.

SC: Can humans become more conscious of their intent?

Grandpa: Yes. We are trying to guide; we are trying to lead. Humans ... are more than they know. Humans ...

SC: Is meditation a method to achieve this?

Grandpa: Yes. Meditation is good to quiet the mind—to ... direct. Yes, meditation is good.

SC: Are there other methods, similar to meditation—or different from— that are useful?

Grandpa: Yes. Pay attention.

SC: Pay attention ... (*Grandpa examines my watch with great interest.*) Do you see on your—on Stan's arm ... do you see his watch?

Grandpa: It is curious. We perceive; we do not see like ... people do. We perceive.

SC: Leo, you have questions also, don't you?

Leo: Yes, if you're willing to respond through Stan to some other questions—they come from other people—that Sarah has written?

Grandpa: Yes.

Leo: Now some of these may be repetitious so ... I'll try to phrase them as she has written them, too.

Grandpa: Please tell Sarah not to be afraid.

Leo: Please tell Sarah ...

Grandpa: Sarah is frightened of her experience, and she will have another one, but it will be very small and very quick. She has nothing to be afraid of.

Leo: OK, good. Thank you. The first question that she has: Do you speak as an individual? Do you speak at a collective level—your group and your race?

Grandpa: Both.

Leo: Both. OK. And she's trying to forestall the possibility that you won't answer a question by saying, "If you answer a question by saying, 'It is not important,' our people—humans—may feel insulted or intimidated. This is what we're trying to avoid as we're learning, but some of us are more advanced than others, and we may be able to help, as we're trying to understand your information." So that's what her hope is—that you will respond to the questions that she has written. So the first question: Can you describe the process of Stan's existence for this "Starseed" or messenger position for the planet? In other words, how do you become Starseed? How do you know that he was one?

Grandpa: He has been chosen. He has been altered genetically. Stan is different. Stan is different from her. Stan is different from conception. He is ... (*slowly*) *refined*. We look for what you call "markers." There are markers that will prove to be more ... *word* ... evolved in certain people, and that's what we look for.

Leo: Then that is how you chose to work with Stan ... these markers?

Grandpa: Yes.

Leo: When you called Leo Sprinkle, you mentioned that permission is granted for more communication. Who grants permission?

Grandpa: Higher. Um … there is a … *word* …

SC: Mind?

Grandpa: No. There is above me as there is below me, like in humans. There are—

Leo: Hierarchy?

Grandpa: Hierarchy, yes. Word, good word, hierarchy. Hierarchy … (*slowly*) *hierarchy*. Yes.

Group: (*Laughter*)

Leo: And are there limitations on our verbal dialogue that limits your—

Grandpa: There is. Humans do not … we do not communicate like humans. Humans cannot … conceive of what we can conceive. They … are slow. They are … slower … it takes effort to slow down to communicate; that's why we use Stan. Stan can communicate and make the communication easier.

Leo: And are there ways that we humans can distinguish your group of Greys from other groups of—

Grandpa: It is instinctive. I have said this before. You guys … guys … guys … Stan's word … guys … strange word, *guys*! Humans have the ability to … decide evil from good. Humans instinctively … know this. Go with … Stan's words … *go with* your instincts. Guys … guys … strange.

Group: (*Laughter*)

Leo: There is a question that Sarah asks about the procedures performed on Stan—difficult for some humans to comprehend and I'm wondering if you wish to add to what you already said about the reason for causing the mental and physical trauma for Stan and his family.

Grandpa: It is not meant to cause trauma, but it is a necessary … situation. Stan is different. Stan must be monitored. If the connection is maintained, he must be monitored. As part of the seven, that is his purpose. He is a leader. He is … *word* … he must be … monitored for his protection in some cases.

Leo: This monitoring—is it done at a distance?

Grandpa: Of course.

Leo: Is there information about how far away the communication or the monitoring occurs?

Grandpa: *Word ... word* ... not important ... but not *allowed*.

Leo: OK. Is there information about the offspring that you produced though Stan?

Grandpa: Yes. They are well. They are also helping ...

Leo: Helping the process of ...

Grandpa: Yes.

Leo: ... human understanding?

Grandpa: Yes.

Leo: Is there information about what they're doing right now?

Grandpa: They are helping. They are ... helping us understand also. Humans are different than we are.

Leo: Thank you. What knowledge do you have about the practice of holistic medicine? Is there something specifically that Stan cannot ingest, that is—

Grandpa: There are things that we will not allow; there are things we will. Stan knows this. There are things that Stan cannot tolerate. Holistic ... (*slowly*) *holistic* ... natural?

Leo: Natural.

Grandpa: Natural is better, but sometimes it does not work.

Leo: Because it's not helpful to Stan's health?

Grandpa: Natural is better because ... but natural is not controlled very well. It is not ... monitored so ... *word* ... some are bad and some are good. Some ... natural are bad and some natural are good.

Leo: So you monitor not only Stan's health but the substances or the medicines that are helpful to him?

Grandpa: We do not want Stan to ... we would prefer Stan well. We do not want ... anything to interfere with ... communication. We do not want anything to interfere with ... the process. It is monitored for Stan.

Leo: OK, thank you. What can Stan or his associates do to become more comfortable with you and your group, knowing that sometimes the visits are traumatic and experimental? What can Stan and his kids do?

Grandpa: Some are not of us.

Leo: OK.

Grandpa: This has been mentioned before.

Leo: Correct.

Grandpa: Some are not of us. Some are ... more traumatic ... some are more traumatic because there are those with ... intentions not like us. There are those *like* us, but not like us ... that have intentions of their own. There is good and bad out there, as there is here.

Leo: M-hmm.

Grandpa: Humans must understand this to understand the situation. We cannot control them. There are those out there that do not want this to happen, just like there are those here who do not want this to happen. There's good and evil everywhere. Evil is a perversion. Evil ... love is a ... universal constant. Love is ... the way it should be.

Leo: Thank you. Sarah asks one last question. Is there anything more that Sarah can do to assist this cause—this case?

Grandpa: To continue what she is doing.

Leo: Thank you very much.

SC: You say there is more than one group interacting with Stan, and some of those other groups don't have the same agenda that you do.

Leo: There's more than one group interacting with humans.

SC: And also with Stan?

Grandpa: Not so much anymore. There are humans and us that are interacting with Stan. There are ... what you'd call ... *word* ... (*slowly*) *hybrids* ... but they are still us.

SC: So most of Stan's encounters have actually been with you ... or your group.

Grandpa: In the beginning, no. Now, yes.

SC: Now yes. Is that because ... why is that so? Why are there not other groups also interacting with Stan?

Grandpa: Some of the groups that were involved in the beginning were not of ... the right intention.

SC: And you were able to defeat them?

Grandpa: Not defeat, but ... persuade them otherwise.

SC: Could you possibly draw ... what you or these other beings look like?

Grandpa: You *know* what we look like.

SC: But can you draw?

Grandpa: There is no reason.

SC: To affirm that what I think is, in fact, what you think?

Grandpa: It is not necessary; Stan has pictures.

SC: Is the physical appearance of ... I think I understand that you can have many physical appearances ...

Grandpa: We can project.

SC: Project. And can you project whatever is comfortable to a particular individual?

Grandpa: If need be. We are physical ... we can project.

SC: Are you always here?

Grandpa: Here ... ?

SC: In our physical space?

Grandpa: Hmm ... we are physical, yes, but we are more advanced. We know of ... things humans do not yet and cannot yet understand. We are ... *word* ... we are more advanced, so we have the ability to ... perceive both physical and nonphysical.

SC: But are you either in physical form or not in physical form? Are you actually in our physical environment?

Grandpa: We are in a physical environment, (*slowly*) *environment,* yes.

SC: And you are visible sometimes but not visible other times. Right?

Grandpa: That is technology that we use. We are physical. There are others more advanced as we are to you that are nonphysical. There are others more advanced to them that are ... we do not yet understand.

SC: Is your concept of time the same as human concept of time?

Grandpa: No.

SC: And is the human concept correct, or not at all correct, or partially correct?

Grandpa: It is partially correct. But it is ... humans have not evolved enough to understand ... or they do not have the ability to understand. They have not yet conquered gravity, let alone ... they have come close. Government, (*slowly*) *government* have come close. But ... they do not truly understand time and/or gravity and its connection with itself. Time is a concept of man. Time is ... a product of motion. Time does not really ... really ... really ... Stan's word ... really ... does not exist the way man think that time exists. Time is ... true time is different.

SC: Thank you. Are there recommendations for me personally—Susan—to encourage my understanding of you?

Grandpa: Yes.

SC: Will you tell me those?

Grandpa: Continue doing what you're doing and ... accept what you see. You have been given gifts, and you must accept them for what they are ... and you must accept the fact that you are who you are. And do not be afraid; fear clouds judgment, and you must explain this to Michael. Michael is fearful but ... he is very strong. Michael is overwhelmed because of fear. And ... fear clouds judgment.

SC: Do you know about his encounter with ... I don't know ... many months ago now ... with fourteen- to sixteen-foot-tall beings?

Grandpa: There are others besides us. There are many that are interested because of Michael's position and Michael's ability. He is visited by many different ... there are many different ...

SC: Can you describe those beings at all to me?

Grandpa: Hmm ... it is not important. Michael has described them to you.

SC: When he is alone, which is often the case, and something that unusual—unusual to humans—occurs, it is understandable that he would be discomforted, at least ... at least temporarily, yes?

Grandpa: Yes. We understand this. He is being ... he is working with others not ... like us but higher than us. He has ... his abilities ... he is a light like Stan is a light. He is ... *word* ... he attracts because his light is so bright, like Stan attracts because his light is so bright. And ... there are others out there that are curious ... just like humans are curious, too ... just like we are curious.

SC: We need to learn more about each other. If we learn more—if we can learn more about each other, I know that this will proceed much faster. The more that you can tell us about yourself, and the more that you can hear about ourselves, the faster we both know that we can trust each other.

Grandpa: I understand, but humans must know that ... we cannot get involved directly ... involved ... we can guide. We have rules just like humans have rules. And these rules have been put in place for the protection of humans and for the protection of us. There is no difference from there to down here. There's good and bad up there, as there's good and bad here. And humans must understand that there is a pro ... protocol ... (*slowly*) *protocol* ... that must be followed ... and that without this protocol there would be chaos. And chaos is a perversion; chaos has led to evil.

SC: Thank you.

Leo: Thank you very much, Grandpa, for the good words.

End of regression session three.

Chapter 4

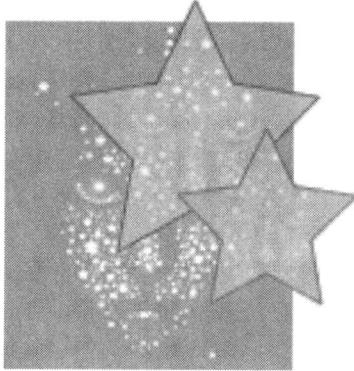

REGRESSION SESSION FOUR

JUNE 21, 2007

The high-level physics equations I had brought forth during sleep and in previous hypnosis sessions caused quite a stir among the physics community. My information on Element 115 even before it had been discovered caused many a chemical engineer and many a physicist to scratch their collective heads. However, prominent MIT and Princeton physicist Dr. Claude Swanson showed continuing interest in pursuing the meaning and the implications of what appeared to be wormhole or black-hole-oriented formulas. I decided to agree to another session with Dr. Sprinkle to see if Dr. Swanson could glean more information on behalf of humankind. I find Claude to be an amazing friend whose openminded-ness has helped me a great deal in grappling with what I originally thought was scrawl. Dr. Swanson has worked at the MIT Science Teaching Center, Brookhaven National Laboratory, and the Virginia cyclotron. Not only has he received a National Science Foundation Fellowship but also a Putnum Fellowship. We are talking about a great scientist here. I wanted to support Claude's efforts as much as he had supported mine, putting his scientific reputation on the line for me with affidavits of authenticity regarding the different scientific formulas.

About three days before the session, Heidi received an Audrey phone call that stressed the importance of sticking to the date of the session. If we did not adhere to this date, it would be quite a while before Grandpa could come through again. In Lisa's forthcoming book, *From My Side of the Bed: Pulling Back the Covers on Extraterrestrial Contact—A Spouse's Point of View*, she explores the possible reason why certain dates had to be arranged for sessions like the one you are about to read. Lisa has examined data that show how certain activities of UFO phenomena cluster around certain times of the year. Her conclusion is that the Orions are using the Orion Stargate to communicate across enormous spans of distance through a wormhole. It's a fascinating read. And frankly, it makes sense to me, especially because we would find out through another session specifically where the Orion Stargate is located. From Claude's analysis, the statements about the Orion Stargate and the wormhole-based formulas I had written down seemed to be connected to Lisa's theory. Claude couldn't wait to find out what new information might come through with this next regression session.

Dr. Leo Sprinkle (Leo): Are you with us, Grandpa?

Stan Romanek (SR/Grandpa): Yes. Hello.

Leo: Hello. Thank you very much for joining Stan. Thank you very much for coming to talk with us. We have many, many questions, and if you are willing, we'll have various people ask questions of you. Is that OK?

Grandpa: Yes.

Leo: Thank you very much. First of all I'd like to introduce Dr. Claude Swanson. He's a physicist, and he has many questions for you about Stan's work and Stan's information. So thank you very much for responding.

Grandpa: Yes.

Claude Swanson (CS): Yes, Grandpa ... I'm Claude Swanson.

Grandpa: We are aware of you.

CS: One of the problems which I'm looking at is ... the equation that Stan writes and is already known. So it would be very helpful to get some kind of result—equation or prediction—that might be verified in the future, that's not yet known, and I wonder if anything of that type might be available?

Grandpa: Yes. We understand. The equations have been given to Stan for ... to qualify ... to verify his experiences. Some are ... more than that. Some are not. Most are to point ... people such as you ... in the direction that needs to be gone. You are the right person for this. You understand, you believe, you ... you'll find the answers. The answers cannot be given. It is ... not allowed. But we can push—we can ... prod in the right direction. Some ... of what Stan has been given is already known. Some ... is not. But it is ... in the right direction.

CS: I'd like to ask you some specific questions about the equations. If you can answer, it would be helpful to our understanding. One of Stan's equations was written backwards.

Grandpa: Yes.

CS: You need a mirror to read it.

Grandpa: Yes.

CS: Can you tell us if any Earth scientist has already published these equations or is about to? That would be helpful for us to trace down if there is someone working with all this.

Grandpa: The theory is known, but ... not in this way. It is ... to my knowledge, not published in this way. But there is information out there about this.

CS: Can you give us a hint on where to look, or what—

Grandpa: Gravity ... (slowly) gravity loop ... (slowly) gravity loop.

CS: Gravity loop?

Grandpa: Stan's word ... loop.

CS: OK. Is it an American scientist?

Grandpa: Hmm ... in America, but not American.

CS: OK. Are those equations—another set of equations that you probably know of relate to [name withheld]'s equations; they're almost the same. I'm sure these equa—

Grandpa: Not [name withheld]'s.

CS: Not?

Grandpa: Someone else's.

CS: OK, let me show you the equation. (*Hands the paper to Dr. Sprinkle, who hands it to me. With eyes closed, Grandpa examines the equation.*) I apologize, I wrote on top of it in red, but the page there is Stan's original equation and ...

Grandpa: (*Handing the equation back*) Yes.

CS: OK. Are you saying they're not [name withheld]'s equations?

Grandpa: No. [Name withheld] ... *word* ... [name withheld] ...

CS: Copied?

Grandpa: Hmm ... yes.

CS: OK. From someone before ...

Grandpa: He has ... added to, but originally they are not [name withheld]. We are aware of [name withheld].

CS: Do those equations relate to the other set that are written backwards? Are they connected, or ...

Grandpa: Everything is connected. Everything that Stan has is interconnected. Everything related to each other—some has ... equations have been given to him over and over again and some not. But everything is interrelated. It is ...*word* ... propulsion. It is pushing in the right direction.

CS: There is a symbol "K" and also the symbol "lambda" in each of the two sets of equations.

Grandpa: Yes.

CS: Are they the same quantities?

Grandpa: I believe so, yes.

CS: One problem with the backward equation is that there seem to be a number of typos—typographical errors. Perhaps a missing equal sign; it appears—

Grandpa: That is Stan. That is ... overload. (*Slowly*) overload. Stan—we were rushed. Stan ... does well, but ... it is hard to communicate long distances. It is hard to ... *word* ... relay information to Stan.

Stan is ... efficient. Stan is ... gifted, but ... sometimes there are gaps.

CS: Would it be possible to get corrections for those equations?

Grandpa: It is not necessary.

CS: Where the equal sign [unintelligible]—

Grandpa: You will figure it out.

CS: (Laughter) Um ... let's see ...

Grandpa: You are smart. Use your instincts, (*slowly*) *instincts* ...

CS: Oh, God. (Laughter) Um ... with the so-called [name withheld] equations—if we call it that—it seems to involve changing the dielectric constants (epsilon) and the permeability (mu) of space. That, that's involving propulsion line in Aramaic?

Grandpa: Yes.

CS: OK. Can you explain physically how you can change those two constants in space?

Grandpa: It is not necessary. This is to guide, to push you in the right direction. When it is time you will figure it out.

CS: Could a beam of charge clusters be effective in helping create such a channel?

Grandpa: That is one way.

CS: The method of propulsion described in Stan's equations appears to involve modifying the speed of light, and the dielectric constant, along the direction to be traveled. Is that true?

Grandpa: Please repeat that.

CS: The method of propulsion that those equations describe appears to involve modifying the speed of light in a particular direction by changing those two constants in that direction. Is that the correct interpretation?

Grandpa: Hmm ... that is one of the interpretations. There is another ... what you call "light" is not ... you humans do not understand light. Light is ... I cannot explain ... you would not understand.

CS: The basic idea in that theory—as Hal Puthoff wrote about in his paper at least—suggests that gravity really is essentially the same as being able to change those two constants, which are mu and epsilon, and that they envelop the dielectric constant. So he keeps saying that changing those two constants basically creates an altered gravity. Is that interpretation correct?

Grandpa: It is ... *word* ... in regards to ... three-dimensional space ... four-dimensional space must play a part in ... these equations. Humans do not yet understand the effects of four-dimensional ... and other-dimensional ... space. Do you understand?

CS: I'm ... I'm not sure. Are the equations basically correct as they stand?

Grandpa: The equations are correct as far as humans can understand, but there is more ... and as ... things progress humans will understand that ... their thinking in three-dimensional space is ... what is holding them back. They must expand to understand that there are more dimensions—what you call, what humans call "dimensions"—involved. And it is not what you expect it is. It is intertwined—everything is intertwined.

CS: Can you give me a hint on what we're missing? Is there a possible way that you can give us a hint about where to look to help understand the higher dimensions?

Grandpa: It is not necessary ... you will figure it out. You are smart. (*Grandpa removes my keys from my pocket and plays with them.*)

CS: Two of Stan's drawings show concentric circles. I'll give you one of them here (*handing drawing to Grandpa*). If you'll just ignore the red ...

Grandpa: Magnetics, magnetics.

CS: OK, that's what I was thinking. In their fields?

Grandpa: Rotating magnetic fields.

CS: OK, good. And is that a depiction of how you create the beam ... which opens the channel?

Grandpa: (*Continuing to play with my keys.*) It is more complicated that that. Um ... energy, energy—high energy, energy that humans do not understand yet. It is ... "Zero point" is as close to understanding as humans can get but ... it takes a lot of energy. (*Gesturing with thumb and forefinger*) touching, touching ... not a straight line, but touching. Humans do not ... humans do not understand. You do not need to travel a straight line. You can (*gesturing*) touch.

CS: But does that diagram ... help us to understand the device or technology?

Grandpa: Yes. It is a ... concept of ... a type of ... power magnetics.

CS: To pull power from space?

Grandpa: Yes.

CS: OK.

Grandpa: Um ... a spinning magnetic field ... can be used as a stabilizer. (*Grandpa closely examines my keys and presses the "lock" button many times.*)

CS: Beside that diagram you have an equation. And it starts off with B sub T, E equals B sub 2P or something, times R squared, and it equals something else, which is under the square root sign? You know that equa—I'll show you (*handing paper to Grandpa*). It's the top equation here on the page that Stan wrote.

Grandpa: Yes.

CS: That appears to be a formula for a magnetic field strength? Is that true, or ...

Grandpa: Yes.

CS: Is that a threshold? What happens? What's the significance of that equation?

Grandpa: It is ... word ... you will figure it out!

CS: (*Laughter*)

Leo: (*Laughter*)

CS: (*Laughing*) We appreciate all help, though.

Leo: (*More laughter*)

Grandpa: Yes, I understand but ... there is ... we cannot give the answers. You must strive to ... humans want ... word ... humans must learn on their own.

CS: May I ask if it refers to extraction of energy from space, or—

Grandpa: In a way, yes.

CS: Or is it more propulsion?

Grandpa: Hmm ... both.

CS: Let's see ... now the units appear to be TM for Tesla meters—is that a correct interpretation of that? The last two letters of that line are "TM"?

Grandpa: Hmm ...

CS: There's a number—1.5 and a minus 18?

Grandpa: You will figure it out.

CS: OK. Is creating an acoustic wave in the same direction that the energy is being transmitted—that's an important thing, is it not?

Grandpa: No.

CS: OK ...

Grandpa: Space is ... absent. Space is a vacuum. Acoustic is sound. There is no sound in space. (*Grandpa finally places my keys down on the end table.*)

CS: In some of the diagrams where you appear to be showing travel, there is a spiral shown down the middle of a channel ...

Grandpa: Yes.

CS: In the middle, there's a spiral. Does the spiral signify that the field is rotated?

Grandpa: That is correct. Very good!

CS: Is there a particular frequency, which is important to use for the energy or the beam going down the channel? Is there a frequency that's important?

Grandpa: Yes. Human's would not understand again—it is ... to do with four-dimensional space than three-dimensional space. It is a ... magnetic frequency. It is a ... power ... much power. To touch ... to touch ... to touch ... it is ... to excite ...

CS: The electrons?

Grandpa: That is correct. There is a ... way to ... small ... very small ...

CS: Small wavelength?

Grandpa: Yes.

CS: A very high frequency—a small wavelength?

Grandpa: Yes, small wavelength.

CS: It is the same frequency or the same wavelength that the electrons vibrate at?

Grandpa: Hmm ... no, it's higher to ... get the electrons excited. (*Picks up my keys and begins playing with them again.*)

CS: OK. Another of Stan's equations has a—(*handing equation to Grandpa*)—at the top left—ignore the red again ... has a V and a delta equals WX over EV—the first—in the upper left part.

Grandpa: Yes.

CS: Is that W energy, or can you tell me a little of what that is referring to?

Grandpa: It is not important. You will figure it out. We cannot give you all the answers; you must strive to find them yourself.

CS: OK … and … I guess I'll stick with these.

Leo: (Laughter)

CS: Stan's son wrote two simple equations on the wall of his room. The first one is an identity for complex numbers. It's simple and well known. The second appears to be a summation theorem involving Bessel functions. What was the purpose of having his son do that— and are there some pieces missing from the second equation?

Grandpa: There are pieces missing, but it is to show Stan that what he is experiencing is real, that … Stan … questions everything. Stan is … skeptical—was skeptical, not anymore. Stan was very skeptical. We wanted to show him that … what he was going through is real. And … his job is very important and very powerful. And that is all.

CS: One of the things I've been trying to work on is a model for physics. There seems to be a connection between particles that are far away from each other.

Grandpa: Yes.

CS: It's what I call the "Synchronized Universe Model."

Grandpa: We are aware.

CS: Is that on the right track, or should I start—

Grandpa: No! You are on the right track!

CS: OK.

Grandpa: That's why you have been chosen.

CS: Can you say anything that will help me …

Leo: (Laughter)

CS: … in developing this theory?

Grandpa: Continue what you are doing. You are on the right track.

CS: OK.

Grandpa: Your theories are sound.

CS: Thank you. I'd like to change the subject a little bit. In the mid-1990s Dr. Courtney Brown wrote a book called *Cosmic Voyage* in which he attempted to remote view a group of ETs that he called "Greys."

Grandpa: (Smiling) M-hmm …

CS: Is there a part of the federation that they can … (laughter)

Grandpa: Humans are easily manipulated.

CS: (*Laughter*) OK. Well, what I was going to ask you is, was it referring to your group and—

Grandpa: There are many groups. Some are ... as they are down here. Some are of good intentions, and some are of their own intentions. Some are of evil intentions ... just as they are down here ... it is the same up there.

CS: Are you saying that most of the information that's in the book is incorrect, then?

Grandpa: Hmm ... I am not familiar with this person, but ... remote viewing ... it is possible, but ... humans have been switched off. Humans have the ability but ... they are ... (*slowly*) *infancy...* infancy ... they are ... Stan's word ... beginners.

CS: In one of Stan's abductions, he was returned wearing a different shirt that we think might be Betty Hill's shirt. Is that correct?

Grandpa: No.

CS: It's a plaid shirt ... with a symbol or something?

Grandpa: No.

CS: It's not?

Grandpa: No.

CS: Why did he get the wrong shirt?

Grandpa: Hmm ... rushed ... rushed ... rushed ... was not Betty Hill's. It was ... rushed.

CS: Was your group in contact with Betty Hill?

Grandpa: Not my group, no. Different race.

CS: Are you expecting that our Earth government might attempt some kind of faked landing at some point in the near future?

Grandpa: We are always concerned about what you call the government. There are ... many groups that are ... against this information. There are many groups that ... are ... it is hard for humans to ... weave through ... the disinformation. But humans have the ability to instinctively ... what humans call "intuition." They know what's right and what's wrong. You must be careful because ... some of the information you have ... might be disinformation. Maybe the government is—what you call the government ... is ... going to try a— what you said—fake landing, but ... that might also be disinformation. We are aware.

Leo: Now, if you will, we ask that you respond to questions from Ben Taylor.

Grandpa: Yes.

Leo: Thank you.

Ben Taylor [pseudonym] (BT): Hello, Grandpa.

Grandpa: Hello.

BT: My name's Ben. Can I ask you a few questions?

Grandpa: Yes.

BT: Did any of the experiences with Stan that I've had—one example being the ship experience, when I was found on a ship with Stan—what was the reason revealed as having significance with your work with Stan?

Grandpa: Yes. Which experience?

BT: On the ship ... with Stan ... on the ship ... when it was recently revealed with Leo Sprinkle and Alejandro Rojas ...

Grandpa: Yes.

BT: Could you, if you don't mind, describe your interaction with the guides that I generally work with and why there appears to be a limited interaction? And also, is there a significance or connection between my spiritual/experiential background—with the entities that I speak of—and Stan? And if so, why?

Grandpa: Your guides are different. Your guides are more advanced. Your guides are ... we all want the same. Um ... we all strive for betterment of man, for ... their ... *word* ... evolutionary ... their ... man has been subdued. Man is in danger of destroying themselves. You are here for the same purpose as Stan, but in a different way. Your guides are more advanced than we are. There is not a lot of interaction because they are more advanced. They are ... angels—you might call them angels; they are more advanced.

BT: What I find interesting is your civilization, your people are also relative, especially more than our society and ... I'm looking for the missing link in why many people on this Earth can interact with these guides and yet why there is what appears to be a limited interaction with you and these entities.

Grandpa: (*Misunderstanding*) There is a limited interaction because humans are not ready. Humans are ... *word* ... hostile. Humans are ...

129

Stan's word—Stan must work on this also—bigot … (*slowly*) *bigoted* … bigoted. Stan knows this. Humans … we do not want humans to become afraid and then aggressive. So we are very cautious. We do not rush in, but things are changing. Humans must grow up. Humans must act quickly or they will destroy themselves. It is a critical point right now. They are at a crossroads and … you also know this. Your guides help you and they try to help Stan, and together we will hopefully find a solution. We are—and Stan knows this—we are … we believe it will happen.

BT: Will you—do you … condone me interacting with Stan or helping Stan? I felt a cautiousness or a … waiting for the right time. Why is that, and do you condone it?

Grandpa: Yes. You are of the seven. You are … a key in the big picture. You are … hesitant … is because Stan is wary. Stan is still skeptical … Stan's mind is trying to grasp what is happening. He is strong, but he has some … skepticism … left. He is cautious. And that's why … there is some hesitancy. You must be involved. It is important that all the keys are together … all the pieces of the puzzle are together … to … help with the enlightenment. Um … your goal is … your *job* is … to help in any way you can. Stan's job is to help in any way *he* can. But … you will help in a different way than Stan. Stan will help in a different way than you. Stan and you and all involved … must be here to make it work. They are … chosen. They are … part of the big picture … as you know … and all of you can feel that … something is happening. What happens depends on … how enlightened humans get … and what they do.

BT: Thank you. Does the psychic explosion and the conversation that I was having with Alejandro a couple months ago have any significance to you?

Grandpa: You are … awakening; Stan is awakening. There are reasons for this. Um … the shift … time restraints … not restraints, but time is critical … things must be … *word* … excited to get … Stan's word— to … get the ball rolling. Everyone involved will … realize abilities they do not understand yet, but … they will … understand … as they happen … the … *word* … the abilities will happen when … it is needed. Do you understand?

BT: Does it have any … that experience—that explosion—does it have any significance directly with you?

Grandpa: With us? No. It is part of the overall process. It has to be with your guides. We were not involved. We were, but … they also understand the importance of … what we are doing. We are doing what they are doing. They also understand that they are at a critical … point. They're … crossroads … crossroads … *word* … crossroads … Humans are at a crossroads … we are trying to point them in the right direction. And humans … are in the physical … who are in the physical, all in the … what you call three-dimensional physical world … is … it's part of the learning experience … even for us. We are more advanced, but there are those that are—as we are to you, more advanced—they are … to us more advanced. There are some that are so advanced that they do not need physical form. And there are some that are even beyond that. We do not understand but … we are learning. But man is … at a critical point. And … he does not understand—man does not understand who they really are. (*Playing with keys again, this time Grandpa places keys on my stomach.*)

BT: Are you familiar with the call or conversations that Alejandro and I had on the phone?

Grandpa: No.

BT: And what is it that I need to do in this … with the entities that I work with?

Grandpa: They will guide you. You [are] doing what you are supposed to do—you must … you and Stan must get together more. You and Stan must … swap ideas. Stan is … on a journey just like you are on a journey; just like everyone is on a journey but … the human race is at a pivotal point. They need assistance and you guys will provide assistance, and you guys will help guide. Stan will help guide through his experiences; you will help guide … with the understanding of … that your guides show you. It is all part of the same thing. There will be seven and … and if they accept, then it will work fine. If they do not accept, then we'll find somebody else to replace them … if possible.

BT: I guess I'm just looking for ... I see a significance in the interconnection between the group that I work with and you as the group with Stan. I guess I'm looking for a deeper significance in that interesting ... I guess "dichotomy" is the word ... I'm looking for the significance in how it relates to you and your group.

Grandpa: We are doing the same thing as your group is doing. Your group is more advanced. We are trying to ... communicate to ... humans that ... they must go a different path. If not, they will destroy themselves. Um ... you are gifted in your ways, and Stan is gifted in his ways. The connection is ... that both of you strive for the same goal. That is the connection.

BT: Why do you answer certain questions that appear to be ... somewhat saying ... with all due respect ... sometimes ... not the actual truth at the time? Is it a matter of *timing*?

Grandpa: It is ... hard for us to communicate. It is hard for us to communicate long distances. We cannot give you all the answers. You must strive for the answers yourself. You know the answer deep down inside. It is redundant to answer the same questions over and over when you know the answer yourself.

BT: So then, it would be accurate to say that ... I was in a conversation with Alejandro and I felt that there was close connection or a communication link that was about to establish with you ... I mentioned the conversation—

Grandpa: We have tried. Your guides are very wary, and it is understandable. Um ... but it is ... we strive for the same thing.

BT: So you were ... you do remember the conversations with Alejandro? Is that correct?

Grandpa: Hmm (*long pause*) ... yes, but we have tried ... we have contacted everybody involved in this group.

BT: Do you have questions for me, sir?

Grandpa: No.

BT: I really appreciate your time.

Leo: Thank you, Ben. Thank you, Grandpa. At this time would you respond to other questions?

Grandpa: Yes.

Leo: We'll ask Sarah Daniels to come and ask her questions. Thank you.

SD: Hi, Grandpa. I want to apologize—I'm a little emotional right now.

Grandpa: That is fine.

SD: It's a little bit stressful to come to this ... and it's so important. Please try to answer each question including some knowledge or reasoning. If you answer questions, "It is not important," our people may feel insulted or intimidated. This is what we're trying to avoid as we're learning about each other.

Grandpa: Susan Carson.

Group: (Laughter)

SD: Some of us are more advanced than others and might be able to help those who are trying to understand your information. Can you do this?

Grandpa: I can try. You must understand ... it is not that we do not want to answer the questions; it is that we are not allowed—like I have explained to Susan Carson.

SD: Correct. It's just important for us to know that it's not allowed, so that we do understand why you are not answering.

Grandpa: I will make it clear!

SD: Thank you. Humans could feel threatened by your contact and experiments upon them—Stan ... I feel it is important to give some explanations for your procedures and to define who you are better.

Grandpa: Not all what Stan has experienced has been from us. Some, yes. There are others involved. There are others involved that are ... not of our group. But it is a necessary thing. Stan must ... word ... Stan must be monitored.

SD: Right.

Grandpa: It is ... in Stan's best interest not to remember, but Stan is stubborn.

SD: OK.

Leo: (Chuckles)

SD: Well, a part of why I want to ask this question is to alleviate a lot of the fear that's going on with humans.

Grandpa: Humans must be wary. There are ... those out there that are here—there are those with good intentions, and there are those with bad intentions. It is OK to be wary, without being paranoid.

SD: What I'm trying to do is gain a trust between your group and humans—to try to help people feel comfortable.

Grandpa: Humans cannot trust their own type.

Group: (*Laughter*)

SD: True. Are you part of a governing association above ... as we discussed?

Grandpa: Yes.

SD: OK. Do you seek friendship and continued guidance after the shift?

Grandpa: Yes.

SD: Do you have a soul?

Grandpa: Hmm ... like humans? It is different, but ... everything continues on.

SD: Has anyone in this room been visited or taken by your group besides Stan?

Grandpa: Yes.

SD: Who?

Grandpa: Lisa.

SD: And?

Grandpa: Heidi.

SD: And?

Grandpa: Paul.

SD: And?

Grandpa: Um ... define "taken."

Group: (*Laughter*)

SD: Well, taken *or* visited—have you been to ...

Grandpa: Everybody has been visited ...

SD: OK, I just wanted to make that clear. Thank you. Have you taken Stan recently ... taken away from his living quarters?

Grandpa: No. It is not necessary.

SD: OK. How did you develop or evolve to become what you are now?

Grandpa: Hmm ... it is (*holding back laughter*) not *important*!

Leo: (Laughter)

SD: OK. Did you evolve on a similar solar system ... like we're in?

Grandpa: Hmm ... humans would not understand. We do not procreate like humans. We do not ... as needed.

SD: If you can time travel, does this mean that you have the ability to change our future, as we know it?

Grandpa: Hmm ... past is hard—I mean, past is *easy*. Future's hard. Future's constantly changing. It is hard ... future's unstable.

SD: Humans experience a wide variety of entertainment. How is your group stimulated, fulfilled, and enlightened?

Grandpa: Humans!

Group: (*Loud laughter*)

SD: Do you feel superior to humans?

Grandpa: Yes.

SD: OK ...

Grandpa: We *are*.

Group: (*More laughter*)

SD: What do you consider to be offensive behavior—from us ...

Grandpa: (*Quietly*) Aggression.

SD: ... if we were face to face with you?

Grandpa: Aggression.

SD: What is an example of aggression to you?

Grandpa: Aggression is ... primitive ... anger ... is primitive. Humans are angry. Humans are easily scared. Humans do not accept their own species ... even though they understand that ... there is little different between each other—because of color or because of ... ethnic background, humans are not very compassionate when it comes to understanding.

SD: This is kind of an important question to me, and I feel very compelled to ask this—that's why I'm asking. Is it possible for human bodies to become another intelligent life form's food supply?

Grandpa: Hmm ... yes.

SD: OK.

Grandpa: But it is not going to happen.

SD: OK. Do you answer to something or follow someone's orders?

Grandpa: Yes.

SD: Who?

Grandpa: It is not allowed.

SD: Are you repeatedly contacting anyone else besides Stan or Stan's team on this planet during this period of time?

Grandpa: No ... well, define "contact."

SD: Contact, meaning ... regular contact, exchange of information ...

Grandpa: We visit, but we do not contact. Stan is ... in my charge.

SD: These questions are to alleviate fear and confusion regarding your contact with Stan and other humans. If these questions remain unanswered, humans will definitely feel threatened. How does Stan have a genetic connection with your group?

Grandpa: Stan is different. Stan is ...

SD: I'm talking about in a bloodline—is there a bloodline connection somewhere ... DNA?

Grandpa: He has been changed genetically; that is all you need to know.

SD: So he wasn't *born* connected?

Grandpa: Hmm ... yes and no.

SD: Tell us what has given you permission to experiment on human beings like Stan.

Grandpa: It is part of the process ... it is not meant to be ... sudden. It is not meant to be violent. But it is kind of a process.

SD: Did you guys have to actually ask for permission?

Grandpa: No, we were ... guided. We were ... like a ... scientist here. We ... were ... given ... orders ... orders ... orders. We have our ... we have our duties like ... humans have duties. There's jobs ... in fact it is more complex ... humans would not understand. It is for a higher purpose. We do not ... *word* ... we can't explain; humans cannot understand.

SD: What were the results of your experiments or testing that were performed on Stan?

Grandpa: He is the right one.

SD: When you were testing our cows, what specifically were you testing for—for environmental purposes? ... Mad Cow Disease?

Grandpa: Hmm ... cows ... we do not ... others, not us.

SD: Is there something that you could say to people who feel that you are trying to deceive?

Grandpa: Deceive ... explain.

SD: Well, some people feel as if you might be trying to deceive or con us for a purpose.

Grandpa: Stan feels just right. It is not our intention to deceive. Humans are very paranoid. Humans are ... paranoid of their own race. They are afraid of what they do not understand. Um ... we try to make as little impact as possible. That's why we do not ... just ... Stan's words ... land on your doorstep. We do not want humans to become fearful or shocked. There are those that do not care, but we are not of those.

SD: Right. And that's why we're trying to get to know you better, so that we can feel more comfortable.

Grandpa: Humans must understand that ... that there is humans, and there is also ... there is also those who are cautious and care. And there are those who do not.

SD: What exactly is your goal or objective with the human race?

Grandpa: The enlightenment ... of the human race.

SD: OK.

Grandpa: To understand that they are not alone.

SD: And so, when we become enlightened, what does that do for you?

Grandpa: It is exciting when any primitive race is enlightened. It is man's turn. We have tried in the past but have failed. And hopefully this time it will not fail.

SD: How do you feel about having a conference for Stan's case ... a conference to get the information out for the public?

Grandpa: It will be. It will be that same as ... the ... Robert is doing it ... documentary ... documentary... it will be. It will all be. It is for ... the benefit. It is to ... show that ... humans are not alone! There is so much more out there, and humans must know this. Humans are ... looking inward.

SD: Just for some encouragement, what type of benefits can we look forward to when joining the galactic neighborhood out in space?

Grandpa: Hmm. You can't conceive ... beyond your comprehension. Humans ... they'll learn so much.

SD: You have said that humans have been guided incorrectly.

Grandpa: Yes.

SD: By whom and what can we do to—

Grandpa: We were ... we are saying that ... humans ... are being guided by ... their own race, but possibly by others ... we are learning. And, if need be, we will deal with it ... when the time comes.

SD: I'm feeling compelled to meet you consciously ... only in a safe, witnessed circumstance, like with possibly others in this group. How can I be more prepared, and can we shake your hand?

Grandpa: Um ... shake hands—human custom.

SD: Right ... it makes us feel comfortable.

Grandpa: When the time is right, it will happen. *When* the time is right.

SD: Do other races also have the ability to create what they imagine?

Grandpa: No. Some do, yes. Humans are gifted this way. Humans do not understand reality ... Humans do not understand time. Humans only believe in three dimensions. Humans barely believe that the world is ... not flat.

Group: *(Laughter)*

SD: Grandpa, what can you tell the team of Stan's case about the release of this information to the public?

Grandpa: There will be those who do not want this out, but do not be discouraged. You are on the right track and ... it will happen, and you must know you are getting help. Everybody that ... everyone involved with Stan is there for a reason. Everyone ... has their part to play, even ... if ... they're not guided by us, but ... guided by ... something more advanced than we are, it is all the same. It is all for learning, and it is all for the enlightenment of man.

SD: Are you aware of my last experience?

Grandpa: Hmm ... most recent or—

SD: Yeah, most recent.

Grandpa: No, I am not.

SD: OK. Is there any information about anything that's *causing* these experiences with me?

Grandpa: You are meant to go through what you go through, just like Stan is meant to go through what he goes through. It is ... a learning experience. It is ... even though you might not understand, it is ... to benefit ... you.

SD: You spoke of the planet that is now the asteroid belt.

Grandpa: Yes.

SD: What happened in the final moments before the destruction of that planet occurred?

Grandpa: They destroyed themselves.

SD: Did your race give any guidance to help prevent the destruction?

Grandpa: We tried.

SD: OK.

Grandpa: They were more advanced than humans.

SD: Oh, they were.

Grandpa: Yes.

SD: Did some of those people have the same Creator, with our makeup?

Grandpa: There is a connection ... genetically with you. Plus ... but ... they were more advanced and ... they just would not listen. Humans are different. Humans ... are ... humans must understand that ... their strengths are spiritual. Their strengths are spiritual, while our strengths are ... technology and ... Earth humans are spiritual. Humans are water vessels and more. Humans ... we are studying this—humans ... are more than they appear.

SD: I would like to ask some questions that might have to do with [unintelligible]. Can you give [unintelligible] in the prehistoric dinosaur period on this planet?

Grandpa: Why?

SD: Well, I think it's a really important part of even *our* evolution ... just the thought that dinosaurs were here.

Grandpa: Yes, they were.

SD: It's very important to me.

Grandpa: There were creatures other than humans. Humans are a part of ... this environment and more. Humans are a conglomeration of ... many things.

SD: But the dinosaurs weren't put here intentionally for ...

Grandpa: No.

SD: ... us humans?

Grandpa: No. They were part of the ecosystem long before humans were developed.

SD: In what path do you perceive to further define quantum gravity?

Grandpa: Humans must be guided correctly. Humans will find their way.

SD: What problem is due to the expansion of the universe?

Grandpa: It is a substance that … not substance, *energy*. Humans are close to understanding it. It is a … *word* … antimatter … antimatter … antimatter … um … best word to describe—Stan's word—antimatter.

SD: We are losing matter going into black holes. What is the point in the black holes we call "the singularity"? And where is all that matter going?

Grandpa: Um … it is … *word* … humans would not understand, again … humans think two-dimensionally. Humans think linearly. You would not understand.

SD: Do you see our planet as an evolving organism?

Grandpa: Yes.

SD: OK.

Grandpa: It is alive.

SD: Do you think we should use stem cells to cure diseases here?

Grandpa: Yes.

Leo: Thank you, Sara. Thank you, Grandpa. Do you think Stan is able to answer more questions?

Grandpa: Hmm … Stan must rest, but … we can come back. Stan's glucose level is dropping somewhat.

Leo: OK, we can come back later. Thank you so much for being with us.

As I come out of hypnosis, I notice the keys on my stomach and ask what they are doing there. Others explain how Grandpa seems fascinated with the buttons. After a period of group interaction, the session is resumed. Leo takes me back into a hypnotic state.

Leo: (*After preparing me*) Are you with us, Grandpa?

Grandpa: Yes.

Leo: Thank you. Thank you for coming, and thank you for your willingness to respond to questions. We had a question from Stan: Is the moon a part of the Earth? Is it an artificial satellite, or is it a satellite that originated with Earth? Do you have information about that question?

Grandpa: Yes. Stan knows this, but … there is … to sustain Earth's life. The Earth … is … in coexistence with the moon, and the moon is … part of the Earth and not part of the Earth. Without the moon, life would not exist on … the Earth. There would be no life. The moon …

regulates ... the Earth. The moon is ... a regulator. And ... without the moon, the Earth would have no breath. The moon is ... was ... in the beginning ... Stan knows this, we told him this ... the moon was a large body that ... struck the Earth ... in the beginning and ... pieces of the Earth and ... pieces of that large body fragmented and congealed into ... what is known as the moon. And without the moon there would be no life. And ... humans are concerned that ... the sun would burn out and life would be extinguished on the planet Earth. But in fact ... the moon is slowly ... drifting away ... from the orbit of the Earth. And once that happens, the Earth's going to lose its breath and ... life will cease. But that is ... way in the future. The Earth is ... because of ... the angle of which this large body struck the Earth ... everything was created. If the angle would have been different, the Earth would, instead of having a moon, it would have a ring like ... some of the bigger planets. It depends on the angle of ... or how the object strikes... the planetary body. The moon is created ... some moons are created ... by large bodies just orbiting the planet, but ... in the case of the moon, the moon was ... a combination of a ... large body and ... combination of pieces ... pieces ... pieces of the Earth congealing with this large body to create a moon. Stan knows this, we told him this.

Leo: Thank you very much. And now some questions from Alejandro Rojas.

Grandpa: Yes.

AR: On the moon, was that large body a natural object?

Grandpa: Yes.

AR: Is there any other large object that passes near the Earth in a great period of time, such as—if you're familiar with the concept of Nibiru ... or a planet that supposedly passes close to the Earth?

Grandpa: Hmm ... there is some ... discussion among humans about ... tenth planet. There are ... humans will see—there are many ... that ... have large ... *word ... word ...* (*makes circular gesture with finger*) ...

AR: Orbits?

Grandpa: Orbits! Thank you—orbits. Large elliptical orbits ... this I think is what you are talking about?

AR: M-hmm. You talked about the sun, and around 2012 will be what we call the "solar maximum"—a very busy part ... solar storm on the sun. Does that play a part in the 2012 date, which was given Stan?

Grandpa: No, not really. It will affect ... all stars have cycles. All stars have ... *word* ... busy cycles and small cycles ... like seasons of a planet. The sun also has these ... seasons itself.

AR: So it doesn't play a role in the date that you gave Stan?

Grandpa: Hmm ... not the sun. Well, it plays a part, but ... not the way you think ... it is ... just ... a natural ... occurrence.

AR: Do you have any more equations or diagrams to show us today?

Grandpa: It is not necessary.

AR: Why was it important to do the regression today rather than at another date?

Grandpa: Communication ... like this is hard. Things must be right ... if things are right, things will not be right for quite some time. If this was to be, it must have been today, or it would not have happened for quite a while.

AR: How long is "quite a while?"

Grandpa: Many months ... maybe a year.

AR: Who sent Stan the laptop? Do you know?

Grandpa: We hope he is getting good use of that.

AR: Yes.

Grandpa: There are ... *word* ... we are in communication with ... it is not important; it is not allowed.

AR: When you answer our questions, it seems you sometimes refer to others to help you answer our questions. Is that true?

Grandpa: Yes.

AR: These others, are they all your race?

Grandpa: Some yes, some no. Some are a combination of ... our race and yours. Some are of a different race. There is a panel ... involved with ... this communication.

AR: Is this a panel that humans are aware of?

Grandpa: No.

AR: There are federations that humans talk about.

Grandpa: Yes.

AR: Is your panel one of these federations?

Grandpa: We are a part of a federation. We are just a small part.

AR: Are humans in contact with some of these federations?

Grandpa: Humans are in contact, yes. That you call ... that you think is government, that you call government ... "federation" is a human word. We do not call it a federation; you would not understand.

AR: Is there a word that is more accurate that we may understand?

Grandpa: It is ... an understanding. It is a ... humans would not understand.

AR: Do you feel unconditional love?

Grandpa: Love is... love is ... *word* ... love is an emotion, but ... love is a constant. Love bonds the universe. Love is ... the *truth*. Love is ... love *is*.

AR: Is love ... what you're trying to describe—some sort of symbiosis?

Grandpa: Humans only understand a small part of what love is. Love is ... love is a connection. Love is ... most, not all, but most intelligent races understand this. Love is ... common sense. Do you understand?

AR: I think I do.

Grandpa: Love is ... *word* ... love is ... the basic binder for everything.

AR: Would Claude be able to find an equation for love?

Grandpa: Hmm ... humans do not understand. Humans ... know the emotion of love. They do not understand the meaning of love.

AR: Do you feel that humans can love your race at the extent that we understand it?

Grandpa: Humans are capable of love, yes. Humans must first learn to love themselves; they do not. Humans are still primitive. When humans can fully understand and love themselves, they will understand their true power.

AR: Do you believe that we're trying to help your efforts?

Grandpa: Yes. We would not be here otherwise.

AR: Many people think it's important to get a conscious consent before taking someone. Is that a possibility?

Grandpa: It is not necessary.

AR: It's difficult for us to explain to others that you're helping, when people are taken without ... and having—

Grandpa: Humans do not understand. Humans are ... strange! Humans are a primitive race. Humans ... won't ... are primitive by design ...

primitive by ... *word* ... humans have been ... guided incorrectly. But humans do not understand. It is not necessary.

AR: Is it part of our goal to help other humans understand your role in the evolution of humanity?

Grandpa: Yes. Your goal is ... I hope that you ... *word* ... enlighten humans. Our goal is to get humans to understand that they are not alone, that there are others out there. Until it happens, humans cannot get the help that they ... so desperately ... want or need. But also, humans ... *word* ... *word* ... humans want ... to learn themselves. Humans want ... Stan has forgotten the word ...

AR: Time?

Grandpa: No. Um ... (*struggling*) I do not understand. Humans ... want to be in charge of themselves ... but ... humans must understand that ... their violence is not going to be allowed. It is not part of the enlightenment. Your job is to ... enlighten them ... by telling that they are not alone, so they can take the focus off themselves.

AR: We bring this issue up a lot because it's difficult for us to reach that goal when there isn't this conscious consent. It's difficult for us humans to accept that.

Grandpa: You are doing fine!

AR: (*Laughter*) A few of us. But it does—

Grandpa: It will not take many to achieve this goal. You all are doing what you are supposed to do. All of you ... have been chosen for a reason.

AR: Ben, earlier, talked about his guides.

Grandpa: Yes.

AR: And you are familiar with them?

Grandpa: Yes.

AR: Do you communicate with them?

Grandpa: Hmm ... we try, but they are guarded. We try, but they are more advanced. They are ... different. They are ... more spiritual than we. We all ... are on the same path. We all ... are striving for the same goal. They understand this, and they are guiding Ben according ... to this also. It is our hope that ... we can all interconnect ... because it is for the same purpose.

AR: Are some of these guides the "creators" that you've referred to?

Grandpa: We do not know yet. They are more advanced.

AR: Do some ... do *humans* communicate with these guides?

Grandpa: They seem to, yes ... unconsciously and ... some have the ability to ... communicate, but ... it is ... by design of ... the ones that are more advanced ... Not by the design of humans, but ... the ones that are more advanced allow it.

AR: Can you name some humans that communicate with these guides?

Grandpa: Yes! Ben!

AR: Are there others?

Leo: There are.

AR: Others you can *name*?

Grandpa: Hmm ... I am not familiar.

AR: Are humans *better* at communicating with these guides than you are ... than your race?

Grandpa: Only by ... the allowance of the ones that are more advanced. They are ... *word* ... *word* ... keep, keep ... keepers. Keepers ... keepers ... strange word ... they are ... keepers ... they ... keepers ... we do not understand yet. We are trying to learn.

AR: Are you trying to learn from us how to communicate with them better?

Grandpa: They are more spiritual than we are. We are ... we are ... advanced in other ways. We understand concepts where humans do not. But there are certain things that we are studying ... from humans and from others. There is importance and even more so ...

AR: Are you trying to learn from us this spirituality?

Grandpa: That is one aspect we are trying to learn, yes. We are spiritual in our own way; humans are spiritual in a different way. Spirituality is ... part of every culture, but ... in different ways. Humans have a gift of it. Humans ... have a strength. Um ... we are ... wondering if humans were made ... for this reason. We are studying it. Fascinating.

AR: Is there a way we can work together in a conscious, deliberate way?

Grandpa: Um ...

AR: ... to help you?

Grandpa: You are doing that now.

AR: Right now, with this discussion?

145

Grandpa: Yes! We are doing that now. This communication is a conscious ...

AR: The last time Stan wrote equations, there were symbols on those equations ... the ancient writing.

Grandpa: Yes.

AR: What were they?

Grandpa: The old language. It was ... to confirm that Stan is ... going through what he is going through. It is to ... verify his experiences, nothing more. We know Stan could not have ... done this. It is to prove to others and to ... mostly to Stan that ... what he is going through is real.

AR: Did you interact with the humans that used that writing?

Grandpa: Hmm ... interact ... with many humans. We know many cultures. We know many ... different backgrounds.

AR: Including that culture?

Grandpa: Yes!

AR: That time period is particularly important to us. Do you know why?

Grandpa: Yes.

AR: Why is that?

Grandpa: Hmm ... a religious—that's why we chose this ... religious ... religion ... religious ... Jesus ... Jesus ... Jesus ... that's writing—it shows that ... because that information will get out to those ... it will be important to those ... that know this ... Jesus and ... it will ... help them to understand ... and become enlightened. It is not by mistake. It is ... for a reason. Everything that has happened ... is for a reason ... down to the smallest ... thing.

AR: So Jesus was someone who existed?

Grandpa: Yes.

AR: Someone who still exists?

Grandpa: Hmm ... there is no doubt ... that Jesus was a ... he was ... we do not understand ... Jesus was not ... us. It was ... part of man's history—part of human history. And ... we are familiar with Jesus. Jesus was real, but ... man has ... confused the message. The message has been confused to ... um ... *word* ... guide incorrectly ... no, no, no ... *word* ...

Paul: Manipulate?

Grandpa: Yes, manipulate. Yes, thank you.

AR: So some people believe they are communicating with Jesus now. Do you know of this?

Grandpa: Hmm ... religion. We are studying. There seems to be a connection with ... those that are more advanced. We are studying. We are studying deity ... deity ... the deity.

AR: Are you trying to establish a better connection with those that are more advanced?

Grandpa: Yes. Always ... to learn.

AR: It seems that humans have a part in manifesting reality.

Grandpa: Yes.

AR: Do you?

Grandpa: Yes. Reality is not ... humans do not understand the connection. Humans do not understand the singularity. Humans are part of ... a singularity. We all are part of a singularity. That is all part ... that is all ... we are all interconnected. We are all ... everything is interconnected.

AR: Some of us have begun to know this.

Grandpa: You know, but you do not understand. Humans do not understand the concept of it.

AR: Do we manifest this reality together?

Grandpa: It is a conscious thing, yes.

AR: Do nonmaterial beings also help manifest this reality?

Grandpa: Yes. It is ... humans would not understand.

AR: But it would seem that we *partially* understand if we are able to ...

Grandpa: Hmm ... yes ... to a small degree, yes, but ... there is so much more. Humans ... have been given the ability to ... consciously ... change their environment, but ... it is more than that. But ... there is a connection ... a singularity—they're all connected. And ... together ... that is what ... manipulates reality. It is more than just physical; it is spiritual. The spiritual ... humans are more than they know. Humans are more than they understand. We are studying this.

AR: Do you put yourself at risk when you visit us?

Grandpa: Yes.

AR: Have humans shot down your craft?

Grandpa: Yes.

AR: Have humans killed some of your people?

Grandpa: Yes.

AR: On behalf of humans, we'd like to apologize.

Grandpa: We understand that not all humans are bad. We understand there are those, just like there is up there, there is down here. We understand there are those that ... mean well, and ... want what's right. We understand that and we appreciate that.

AR: These men ... do you fear that?

Grandpa: Hmm ... there is no doubt. Just change.

AR: Just a couple more things: About the seven—

Grandpa: Yes.

AR: You said previously that Susan *knew* of the seven,

Grandpa: Yes.

AR: ... but was not *of* the seven.

Grandpa: That's because we were not sure she was ready. If she does not want to be part of the seven, we will try to find somebody else. But ... it is hard for us to ... find ... there are certain ... characteristics we look for ... to help. Everyone must play a part. Everyone must be open ... the seven ... of the seven must be open ... to the experience and must be open to ... help ... um ... *word* ... guide, and help mentor ... the population. It is important ... that they do this ... without scaring or that ... overloading... Stan has been chosen because he is very good at this. He understands. He has been on both sides.

AR: Previously you also said Susan *knew* of the seven?

Grandpa: She does. She does know the seven, and deep down inside she knows ... that she is part of the seven. She does not want to accept it, though. Even now she is questioning.

AR: As far as I know, she doesn't know Ben.

Grandpa: No, there's one ... she does not know of. We know this. She knows this. She knows of the others. If she's not willing to be part of the seven. We will try to find somebody that ... will be willing. Susan will be part of the seven because she is good at what she does. She is ... she can bring together. She is ... a binder. She is a binder ... binding agent ... of truth ... *word* ... binder.

AR: Earlier you said that you take orders.

Grandpa: Yes.

AR: And you said, like a scientist here.

Grandpa: Yes. We have ... a purpose. There is a reason why we're here. We do it ... because it is *right*.

AR: Who was the scientist you are referring to?

Grandpa: Um ... *word* ... a ... scientist ... just any scientist. We were using it—Stan's word—as an example ... *word* ... just to ... explain ... and it is because ... that is what we're supposed to do. Humans are lost. Humans ... are stuck in this place ... trying to find themselves. We know what we are supposed to do. It is instinctive.

AR: Are your people still visiting [Stan's sister]?

Grandpa: No.

AR: Are others that you know of?

Grandpa: That I know of? No. [Stan's sister] is ill.

AR: Are you aware of what happened to me last night?

Grandpa: No.

AR: All right, that's all.

Leo: Thank you, Grandpa. Thank you, Alejandro. *(Grandpa begins to play with my sunglasses.)*

PW: Hello, Grandpa. As you know, I am Paul—a friend of Stan's ...

Grandpa: Yes, Paul.

PW: And I'm helping Stan with his book.

Grandpa: Yes, Paul.

PW: You once mentioned that some of Stan's friends need to be aware that paranoia is not good.

Grandpa: Yes, Paul.

PW: Can you please tell me what it is specifically about me that inspired this warning? Is it because I sometimes assume the worst in a situation ...

Grandpa: No, Paul.

PW: ... or is it due to my investigation into the dark forces, which are misguiding humanity?

Grandpa: Yes, there are dark forces, but there are also good forces. Um ... paranoia ... to ... solely concentrate on just the dark forces is ... negative and is ... self-destructive. It is ... OK to be cautious ... but it is not good to be paranoid. It is OK to ... seek the truth, but it is ...

not good ... to concentrate on ... just the bad. There is good; there is bad. And yes, right now ... bad seems to outweigh the good ... as far as the human race is concerned, but ... (*slowly*) *Stan's words* ... the pendulum swings both ways.

PW: Well, I'm confident that the good will win over the bad, but it is a fact that there is a criminal element in control of governments, in control of the currency, in control of universities, in control of the media ...

Grandpa: We understand this. We know this. Just—

PW: So it's a fact; it's true then.

Grandpa: It is true, but ... to concentrate solely on this is unhealthy. It is unhealthy from ... it is unhealthy for humans mentally. It's good to be wary, but ... to a point it becomes an obsession ... and you cannot concentrate on ... what your real goals should be ... and what your gifts are.

PW: I always felt that one goal should be exposing the enemy so that humans can overcome—

Grandpa: They will expose *themselves*; it is the nature of the beast. They will make a mistake. It is OK to seek the truth. It is OK to enlighten. To be paranoid is not OK ... because it takes away ... from ... what your true gifts are.

PW: Thank you. Do you have any awareness regarding my trip to Brazil in 2004, or does your awareness and monitoring of me personally begin after that event?

Grandpa: After.

PW: OK.

Grandpa: And we do not monitor all the time; there is no need. That would be a waste.

PW: Do you enjoy human art and music?

Grandpa: Hmm ... it is fascinating. We are different. What we enjoy is different. We find things fascinating ... humans are a diverse race. They can do many things. Some of it to us seems silly; some of it does not. Humans ... (*giggles*) sometimes are entertaining.

PW: Thank you. I'm going to give someone else the opportunity to ask questions. Thank you very much (*leaving seat*).

Leo: Thank you, Paul. Thank you, Grandpa.

Grandpa: (*Grandpa calls Paul back*) Paul!

PW: (*Returning to seat*) Yes, sir?

Grandpa: You are smart.

PW: Thank you!

Group: (*Laughter*)

Grandpa: But ... you are missing the point. You must relax. You have abilities you do not yet know of, but ... you are clouding your own judgment. That is all.

PW: Thank you. I'd like to be an effective tool for the transformation and the shift.

Grandpa: Stop being paranoid and you will be.

PW: OK, thank you.

Group: (*Laughter*)

H: Hi, Grandpa. I'm Heidi. (*Grandpa examines my sunglasses again.*)

Grandpa: Yes! I know of you!

H: I was wondering if the importance of us meeting today rather than later has anything to do with the fact that it's the summer solstice?

Grandpa: Yes.

H: And why? How does that tie in?

Grandpa: It is easier to communicate ... and it's hard to explain. Things must be aligned ... energies must be aligned in order to communicate. Um ... right now we are long distance—what *you* call long distances ... we use a technology humans cannot understand yet. But things must be right. And ... will not be right again for quite some time. If this was not to happen, we would not have communication for some time.

H: Does that include phone calls from who we call "Audrey"?

Grandpa: Yes.

H: So we should not expect to hear anything for—

Grandpa: No. We will monitor you if need be. We will ... we cannot communicate long distances, but if need be we will come closer.

H: Last night, Stan saw two craft near the moon. Does that have anything ...

Grandpa: Yes.

H: ... to do with your visit tonight?

Grandpa: Yes ... it was our way to say "hello."

H: On another subject, do those of your race—we kind of touched on this earlier ... on love: Do you—

Grandpa: Love is a universal constant. It's what binds things together.

H: M-hmm. But—

Grandpa: The conscious universe was made ... with love. We are studying this.

H: Are you a compassionate being?

Grandpa: Hmm ... define "compassion."

H: Caring deeply for others.

Grandpa: Humans are emotional. We do not have emotions like humans. We ... try to ... we do not want to hurt or to ... frighten, or to ... so yes, we are compassionate.

H: You had said earlier that I had been taken.

Grandpa: Yes.

H: How many times?

Grandpa: Twice.

H: Were those times the two dreams? Was one of those times—?

Grandpa: Yes.

H: Do I have an implant in my ear?

Grandpa: It is not important.

H: OK. I would *love* to gain a memory of what happened, or be taken again and have a memory of it. Will that happen?

Grandpa: (*Quietly*) Strange ...

Group: (*Laughter*)

H: Why is that strange?

Grandpa: Humans do not want to be taken.

H: I think it's fascinating!

Grandpa: If it is necessary. Right now it is not necessary. You are doing what you are supposed to be. You are a good support ... for Stan and Lisa and ... you are doing what you are supposed to be doing. You are ... involved with Stan and Lisa and ... you play a role ... as part of the group. And we appreciate what you have done for Stan and Lisa.

H: You said, when [Paul] asked, if you enjoy our art and music, you said you enjoy other things ...

Grandpa: We do enjoy ... hard to explain—we are different. We are different; we perceive differently and we do not think like humans.

H: Well, what do you enjoy?

Grandpa: Hmm ... it is not necessary ... *humans.*

Group: (*Laughter*)

H: *Aside* from humans, I mean. Do you have your own kind of music and art?

Grandpa: We have our own ... type of entertainment, yes. It is beyond humans. Humans ... cannot confuse. Humans ... are still primitive. Humans can only think linearly. Um ... we do not communicate like humans—it is too slow. Humans would not understand.

H: You find us interesting, huh?

Grandpa: Yes.

H: OK, that's all I have. Thank you. (*Heidi is handed some handwritten questions from Lisa, who did not wish to appear on camera.*)

H: Oh, OK. (*Reading from Lisa's questions*) Are you aware of Stan and Lisa's new home?

Grandpa: Yes.

H: Is this a positive place for them?

Grandpa: Definitely.

H: Any problems with the home that they should be aware of?

Grandpa: No. They have chosen wisely.

H: Have you been to the home?

Grandpa: Yes.

H: Did you make the darts disappear last weekend when Stan and Heidi were playing?

Grandpa: (*Coyly*) It is not important.

Group: (*Laughter*)

H: How about turning on the water faucets today?

Grandpa: (*Smiling*) It is not important!

Group: (*Raucous laughter*)

H: No fair, Grandpa.

Group: (*Laughter*)

H: Stan and Lisa felt ... that—after seeing the two craft last night—that you were there. And you said you were?

Grandpa: It is not important.

H: Why was there a feeling of anger between the two of them at 2:00 AM?

Grandpa: We do not know.

H: Why were the two medicines burnt—that Stan had been prescribed?

Grandpa: Stan should not take them. It is not good for Stan.

H: If he is not supposed to take those medicines, why don't you fix what the medicines were to treat?

Grandpa: It will fix itself. Stan ... we are still ... we do not understand all of what Stan is going through. But we are learning. Those medicines would harm Stan. Stan is different. He does not metabolize like normal humans do because Stan is not ... like a normal human.

H: What part does each of us in this room play? And please address each of us in what we need to do. And please do not say we are doing what we are supposed to be doing.

Leo: (*Laughter*)

Grandpa: Um ... it is ... *word* ... to guide, to support ... everyone plays an important role. It would take too long to explain each and everyone's role in this, but ... they are doing what they are supposed to be doing! They are ... supporting Stan. They are ... getting the information out where it needs to get ... they are ... there when Stan and Lisa need them. They are a support system. Stan and Lisa's mission would not be able to do what they are doing. *Stan* would not be able to do what he is doing without the support system. He would not be able to do what he is doing without Lisa. He would not be able to do what he is doing without ... any of the people involved because he would not have a support system.

H: We don't mean to be redundant in asking ... (*Grandpa holds up his finger as if requesting a pause.*)

Grandpa: (*Pause*) OK.

H: We don't mean to be redundant in asking that question. I think that each of us just wants to be sure we're doing everything possible.

Grandpa: I understand. I would let you know if you weren't, but everyone is doing what they are supposed to be doing, and they should be commended on what they are doing. It will be harder as time goes on because there are those that do not want this out. So just be prepared.

H: Does your main enjoyment come from learning or from just watching humans?

Grandpa: Learning ... learning ... learning things. Every intelligent life form strives to learn.

H: OK. That's it—thank you!

Leo: Thank you. We have some questions from Rick Nelson.

RN: Thank you, Grandpa and Stan. This is a fascinating experience for all of us. It appears you are a more intellectual, scientific, and technological race than us. You have indicated that we have more potential ... we are more spiritual. (*Grandpa puts on my sunglasses and then removes them.*)

Grandpa: Yes.

RN: Could you elaborate on that—our potential—what abilities we have that we underestimate about humans?

Grandpa: Humans do not understand who they really are. Humans are a conglomeration of many things. Humans have been guided incorrectly. Humans ... have been turned off. We are studying this. We do not understand this. Humans ... have the ability, as we discussed before, to ... consciously manipulate their environment ... as a whole. Humans can achieve great things once they realize that they must... stick together to do this. Um ... Stan's words ... by themselves, they are powerful ... together, they are magnificent.

RN: Thank you. Are there humans that are not entirely human around Stan?

Grandpa: Yes.

RN: What percentage would you say?

Grandpa: Hmm ... it depends. Stan is ... small percent.

RN: Are there those in this room that are not entirely human?

Grandpa: Yes.

RN: (*Laughing*) I'm afraid to ask ...

Group: (*Laughter*)

RN: Am I one?

Leo: We all are.

Grandpa: "Human" is a word you use, but ... to answer your question—you are a conglomeration of many things, a little more than out there. It's mostly out there. Stan is different, though, he has ... been altered.

RN: Have I been taken?

Grandpa: Yes. Not by us, but by others.

RN: How neat! Can we define a black hole?

Grandpa: It is a singularity. It is a oneness ... it is hard to explain. Humans would not understand. Um ... it is a ... *word* ... it is a crux of matter ... it is a ... it is hard to explain.

RN: Thank you. Orbs—what we refer to as orbs are nonconventional.

Grandpa: Yes. We have discussed this before.

RN: Did you use orbs for travel?

Grandpa: No.

RN: For observation?

Grandpa: You would not understand. Um ... observation—there are orbs Stan has seen—those are ... observation. Orbs you are talking about ... are ... some are natural. Some are ... not ... some are ... entities ... some are *you*.

RN: Would you or others intervene or stop an asteroid if it were to seriously harm the Earth?

Grandpa: We would not be allowed. It is ... humans have the ability. Humans are more than they know. Humans are ... we are here to guide. Humans think that ... if a thing happens it is evil or it is good. But in a lot of cases, it is ... what it is. It is not ... it is neither evil or good. It is ... what it is. If ... a large body ... hits the Earth ... it would be to the humans to defend themselves. That was part of the test.

RN: When we have UFO or paranormal gatherings or events or expeditions, we have found it quite beneficial for acceptance when we have interaction from you or others like you.

Grandpa: Yes.

RN: Would you be willing to continue or step up some of this exp—

Grandpa: It will happen. But ... everyone... must ... be willing. There are those that say they're willing but are ... terrified. We will not ... we will not ... terrify. We will not scare.

RN: Are there things to share with Stan that you'd like to—that would be helpful to him?

Grandpa: Hmm ... to Stan and Lisa both—they need to stop fighting. We understand the stress, but ... they will get through it; they will be fine. They need to ... start relying on each other.

RN: Would you be willing to visit tonight, later?

Grandpa: It is undecided.

RN: We'd appreciate it. Thank you. How many dimensions are there?

Grandpa: You will learn. There are many.

RN: How many races, to your knowledge?

Grandpa: We do not even know. There are many.

RN: Thank you. That's all my questions. I appreciate it. (*Grandpa puts on my sunglasses again.*)

Leo: Thank you. There's just a few minutes left on the tape. Grandpa, do you think that Stan is tiring?

Grandpa: Yes. He needs a break.

Leo: OK. Thank you very much for answering these questions. We appreciate it.

Grandpa: Yes.

Leo: Thank you.

Once again, a break is taken so I can regain my energy. The group exchanges thoughts over what has been discussed, and what more needs to be discussed. The session begins again.

Leo: Greetings. Are you with us, Grandpa?

Grandpa: Yes.

Leo: Thank you for being here, and thank you very much for responding to questions. We have some questions now for you.

Grandpa: Yes.

Richard Summers (RS): Hello Grandpa. This is Richard.

Grandpa: Yes.

RS: I just have a few questions here. It's come to my understanding that Stan has children with you ... on your ship, possibly?

Grandpa: (*Silent, with a confused expression.*)

RS: Is that true?

Grandpa: Not with *me* ...

RS: Not with you *specifically* ... but with your race?

Grandpa: Yes.

RS: What is the goal you wish to achieve with this interbreeding?

Grandpa: Understanding.

RS: Understanding of what?

Grandpa: We are learning. We do not ... quite understand you yet. They are more ... they seem to be more ... they are ... more than we understand, and we are learning.

RS: Now, it was brought up before about the scenario of an asteroid coming to the Earth.

Grandpa: Yes.

RS: And we would know what to do if that [situation] did arise?

Grandpa: Yes.

RS: This is a two-part question: Did this happen before—did an asteroid visit the Earth and mankind was destroyed?

Grandpa: Not mankind, no. But ... yes, it has happened before.

RS: Now, if we would know what to do, is that any relation to our unconscious mind?

Grandpa: No.

RS: No?

Grandpa: Humans have the technology to ... save themselves if need be. But that is not a big concern right now.

RS: OK. When you mentioned earlier that you only procreate as needed ...

Grandpa: Yes.

RS: Does this have to do with a balance in your ecology?

Grandpa: Yes. We do not ... humans are taxing the Earth. They are taxing the system in which they live. We do not do that. We ... create as needed. We do not overwhelm the environment we are in.

RS: Question—I guess this is for Heidi: Leo's spirit guides told Heidi that she was an ET in a past life. Was there any connection to you or your group in that life?

Grandpa: I am not aware.

RS: Now, when you mentioned earlier about the interbreeding and that you're curious—does this have anything to do with ... let me back up a bit. It's known to us that in our conscious mind that we are—aware of—only ten percent aware of our thoughts?

Grandpa: Yes.

RS: Ninety percent lies in the unconscious.

Grandpa: That is not true either.

RS: Can you elaborate on that?

Grandpa: Part of it is in the unconscious, and part of it in conscious, but some has been turned off. We are studying this. We did not do it. We do not quite know who did.

RS: When you mention "turn off," is that sort of a blocking of ...

Grandpa: Yes ... yes.

RS: ... of why we have so *much* in our unconscious, as opposed to being in the conscious?

Grandpa: Yes. There is more to humans. When the shift happens, we believe this will just automatically turn on.

RS: OK. I understand completely.

Grandpa: Humans are more than they understand. Humans are much more than they understand.

RS: One last question: We are all going over to Stan and Lisa's tonight. Can you come by?

Grandpa: It depends!

Group: (*Laughter*)

RS: On?

Grandpa: It just depends!

RS: OK. Thank you very much. That's it for me.

Leo: Thank you, Richard.

RS: Thank you.

Leo: Thank you, Grandpa. (*John comes forward with no introduction.*)

Grandpa: Hello, John.

John: Is there anything that we can avail ourselves of—any processes that we aren't using—to bring the information and the messages that you're attempting to help us with more effectively ... to the people?

Grandpa: Hmm ... knowledge is power. Stan knows this.

John: Is there anything that we are missing? Are there any avenues that we should go down that we haven't been exploring?

Grandpa: Humans are on the right—everyone who is in this group is on the right track, but ... it would be beneficial to look at it from both sides. Stan has ... a gift do to with Stan. Those on ... those who have been on both sides as ... what you call a "skeptic" and as ... a believer—you must think as a skeptic to understand ... how better to get this out. Because there are many who don't believe, and many who are afraid to believe. And it is about fear. And you must understand that it is about fear. And you must ... come up with ways to ... mother their fears.

John: Would you suggest that it would be an efficient ... means for us to utilize the political process that exists today ...

Grandpa: Yes.

John: ... to accomplish some form of change?

Grandpa: This information is for everybody. Politicians and ... the corrupt and ... this is for everybody. Um ... we understand that those who are corrupt and ... those who are ... in it for their own gain are ... *lost,* and they need to be enlightened ... like all ... on this planet.

John: Do you have any specific suggestions on how this might be accomplished, how—most don't want to hear that.

Grandpa: What ... the group is doing now is a first step, from there it will snowball. And ... everyone here must know that they're being helped outside of ... just this group. There is a ... it is ... an inner ... *word* ... there is help ... from above also.

John: Yes. That's helpful, just knowing that. Thank you. When the shift happens, should we expect some physiological changes, for example in our ... I would say ... the ... chemical soup ... that is secreted within the pituitary and our endocrine system ... will that have an effect on the way we think?

Grandpa: It is mostly ... *word* ... consciously ... Humans will not grow a third arm or a third eye or ...

Group: (*Laughter*)

Grandpa: ... but ... they will grow ... to understand. They will ... be enlightened. They will ... discover powers they do not know they have.

John: Is there a physiological component to this change in the way it's—

Grandpa: As far as your mind is concerned, yes. As far as the body's concerned, not so much, no.

John: That was my question. I don't mean to be self-serving, but would you have any specific suggestions for myself on how I can be more effective?

Grandpa: Hmm ... supporting Stan and Lisa is ... effective. Stan is ... what Stan is having to deal with has been hard and ... he has a ... great responsibility. He must have support in order to do this correctly. By supporting Stan, you are on the right track. Stan is not the only one, though. There are six others ... that is ... seven ... and when it becomes time ... all seven need to be supported. It is going to be tricky to get them together, though. There is a problem of distance and of travel and of ... communication. Your world is fractured into multiple ... linguistic categories and ... there will need to be interpreters and ... there will need to be a way to get them together. Financially ... it is going to be interesting to see how this is going to be ... we will try to help, but there's only so much we can do. It will be left up to the people involved to try to help.

John: Thank you. (*Claude Swanson approaches.*)

CS: I just had a quick question, Grandpa.

Grandpa: Yes!

CS: Thanks for your help earlier, by the way, in the physics questions.

Grandpa: Yes!

CS: In one of your previous sessions, you mentioned something like fifty percent of the people on the planet, if they recognize the existence of ETs or something, then you have permission to come over and—

Grandpa: The shift will automatically happen. It will have nothing to do with us. It will automatically happen. It is ... an enlightenment. It is a consciousness issue ... and as we discussed earlier, humans have the ability to change their surroundings ... consciously. But it has to be done together; it cannot be done singularly or one person at a time. It has to be done together. This also holds true for the shift to happen. They must believe that they are not alone. Fifty percent of the world ... just even a little bit more than fifty percent of the world. Then the shift will happen.

CS: The trick is to get a gradual change.

Grandpa: It could be gradual, but we … believe it will be fast and … glorious.

CS: Thank you, Grandpa. (*Ben Taylor approaches.*)

BT: Hello, Grandpa.

Grandpa: Yes!

BT: Do you mind of I ask you one more question?

Grandpa: (*Shaking head*) yes, I do.

Group: (*Laughter*)

Grandpa: I do not mind.

BT: Grandpa, this has been something that has been on my mind, and it's … (*Grandpa indicates a need for momentary pause.*)

Grandpa: (*Long pause*) OK.

BT: It's … I want to get this possible clarity on … the episode that I experienced with Stan. One of the episodes involved … a lot of very, very difficult—I would say emotional situations with Stan that … I was nurturing him on … that ship that I described earlier. How is that affiliated with your group and—I just was hoping for some clarity on how that is affiliated with you and your project.

Grandpa: Your guides … have the same goal as we … you are connected to Stan in some way we do not yet understand. We do not know if it's by design … from your guides … or if it is … what it is. There seems to be a connection. We believe that … of all we think, along with others, we have orchestrated—the gathering—we now wonder if it's been orchestrated by something greater than us. And as time goes—what you call "time"—goes by … Stan's words … it seems to be … that we are not … in as much control as we thought … that this is being orchestrated by something greater than us.

BT: Thank you very much, Grandpa. (*Sarah Daniels approaches.*)

SD: Hello again. I had some questions that I'd like to go over real quick before we end the session. Would you like me to ask you these questions a little bit slower than I did last time?

Grandpa: No.

SD: OK.

Grandpa: If you ask the same questions, they will get the same answers.

Group: (*Laughter*)

SD: No, very different, very different. Do we all ... does everyone here in this group have an implant?

Grandpa: No.

SD: Can you give me an explanation of the experiences that I've had ... that are extra-dimensional related?

Grandpa: It is not allowed.

SD: Why did Stan have elemental bismuth when he came back to his home? Why was that in your craft?

Grandpa: It is a byproduct.

SD: Was it something that you had picked up from another place?

Grandpa: No. It is a byproduct.

SD: A byproduct. Can you explain that to me?

Grandpa: No.

SD: OK.

Grandpa: It is not allowed.

SD: OK. Thank you. One night I was in my room—I can't really give you a particular time, but I was in my room, and I felt like I couldn't move. Have you guys ever been to my room?

Grandpa: We have, yes. But the experience you are talking about is a biological experience.

SD: Biological? OK.

Grandpa: It is not from us.

SD: Have you visited my children?

Grandpa: Hmm ... in passing.

SD: My oldest daughter speaks of experiences where she's aware of a lot of different races. Is this something that she's making up, or is she actually having contact?

Grandpa: Hmm ... I do not know.

SD: Do you use polyvinylpyrrolidone for in-vitro fertilization?

Grandpa: Hmmm ... we use ... not—that is a human thing. We use something like that, but different.

SD: What you are referring to is the—

Grandpa: That was found on the nightgown Stan came back with, yes.

SD: OK. Are you aware of any special abilities that I have?

Grandpa: Yes. You are empathic. But ... you are a beginner.

SD: Is the implant in Stan a manufactured, biological object?

Grandpa: It is not important. It is not allowed.

SD: OK. And I imagine that you couldn't tell me who created the object.

Grandpa: We did.

SD: Is it possible that we could get some kind of a demonstration from you this evening ... for credibility purposes?

Grandpa: It is not important. That is asked every time, but ... no.

SD: Why is it that you arrive so late at night pretty consistently?

Grandpa: It is a ... matter of ... energy. Displacement and placement and ... it is a matter of ... alignment.

SD: With the planets?

Grandpa: It is more complicated that that.

SD: During one of my experiences, I had a death threat. You are probably not aware of that?

Grandpa: No.

SD: OK, so you were not involved. It was a pretty mean, pretty scary one.

Grandpa: There are ... those who do not want ... the enlightenment to happen and ... those that are ... like us, and/or of us and are even human, that ... know that this is taking place and ... instead of taking place, they are threatened by it.

SD: Yes, I was threatened quite a bit, actually.

Grandpa: Yes. Stan has also. And others may be in the future, but ... we [unintelligible].

SD: I notice that you're using math a lot with Stan. Is that kind of the universal language?

Grandpa: Hmm ... concepts and math, yes. Humans know this and ... mostly to qualify Stan's experiences. Stan's weakness is in this and ... we want to prove Stan's experiences are real, so ... we give them something that they know Stan is unable to do ... to prove that Stan's experiences are real.

SD: Can I visit you at Stan and Lisa's home?

Grandpa: We are not *at* Stan and Lisa's home.

Leo: (*Laughter*)

SD: I'm sorry. Can we ... or team members here—can they visit with your group at Stan and Lisa's home?

Grandpa: If need be ... and in time. Humans are not ready. I know it is important for humans to understand. This group will experience things, but ... not now. In small amounts ... in what you call "sightings" and ... maybe a brief visit, but ... that is all that is allowed.

SD: OK, all right. Is there anything we can do to remove the block that our minds are shut down with?

Grandpa: No.

SD: OK. So right now what we have for the seven is Ben, Stan, Susan, Michael, ... and then, um ...

Grandpa: *Maybe* Susan.

SD: Maybe Susan ... OK. So there possibly can be three to four others.

Grandpa: Susan knows of them. But it is up to the individual if they want to be a part. We do not want to force them. They have to come willingly. They have to accept their ... their duties. They have to understand that this is not ... easy. But it is a great responsibility. And ... that it serves the same purpose, but it needs to be ... all the pieces to complete the puzzle. And without the pieces, the puzzle will not be completed ... and will not serve as ... will not be effective.

SD: Well, we're all looking forward to bonding with you some more.

Grandpa: Bonding ...

SD: We love the fact that you lend your assistance, and we hope for more contact.

Grandpa: There will be more contact, but it is hard for us to ... communicate this way. It will be some time before we can do it again, and we are ... going to be out of phase ... for a while.

SD: How is it that you're out of phase?

Grandpa: Humans do not understand. Things must be ... in correct alignment, and they will be out of alignment.

RS: Is there another means of communication?

Grandpa: Unless we come closer, no.

RS: Will you come closer?

Grandpa: It depends.

RS: On?

Grandpa: It *depends*.

H: On?

Group: (*Laughter*)

Grandpa: There is no "on" … just, it depends.

RS: Thank you.

SD: For a little more self-confidence, what type of protection do we have in the things that we need to be involved in with this case?

Grandpa: It is not important; just know that you are protected. There is only so much we can do, but you have lots of protection. I will allow one more answer. Stan is getting fatigued.

Leo: Thank you.

AR: OK, for the last question, you just told Sarah that the substance that was on the clothing Stan came back with was something you do not use?

Grandpa: No, they … they … analyzing it, they are missing part of … it is similar, but it is not used for what you think it is used for.

AR: What is it used for?

Grandpa: Hmm … it is not allowed. That was an accident. Stan was rushed and … that is all.

AR: All right, thank you.

Leo: Thank you, Alejandro. Thank you, Grandpa, for being with us and for helping Stan and the group continue their work. Thank you so very much.

Grandpa: Yes.

Leo: We appreciate it. (*Grandpa indicates he has something more to say by raising a finger.*)

Leo: Yes?

Grandpa: Everyone here is doing what they are supposed to. Continue. That is all.

Leo: Thank you. Good-bye.

End of regression session four.

Chapter 5

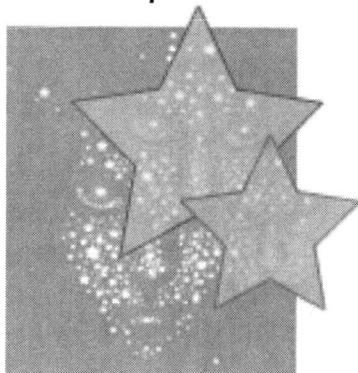

REGRESSION SESSION FIVE

AUGUST 23, 2008

The previous regressions sessions had produced a library of answers. But they caused even more questions to arise. My inner circle of friends began to look at themselves differently. They also found themselves looking at the world differently. If humanity truly was on the brink of termination, what role could they play in getting our world to wake up and pay attention? One of the dangers of any small group that begins to see itself in a position of telling the world what to do is that of projecting its own belief systems and views upon others. Many discussions arose around this. What came to light was our recognition that we could not bring change in the world by thinking of ourselves or our world in the same old ways that had gotten us into this mess in the first place. The hidden implications and sidebar comments from Grandpa suggested that humanity was far more than it realized. How were we to find out what that meant? There was only one way: We had to look within. We had to fully examine what blind spots we had imposed upon ourselves. We had to take the time and effort to look at ourselves. For it is through each one of us that the quantum world actually functions. The equations implied that, as did Grandpa himself.

On August 11, 2007, four of our group decided to spend a day at Robert's cabin. Both Heidi and I ended up being abducted. Neither of us had any memory of it. Robert found a trampled twelve-foot circle in the field with three landing pod marks inside it.

Over the next year several unexplainable events occurred around our house, including an attempt by a Black Ops to shoot me. Death threats were made toward my inner circle of friends. We felt it was time to get information as to how the bigger picture might connect with all the activity we had witnessed. We sought out Dr. Leo Sprinkle once more.

Dr. Leo Sprinkle (Leo): Any information from Stan's memory regarding August 11, 2008? [Leo had gotten the year wrong. It was 2007.]

Stan Romanek (SR/Grandpa): (*Long silence*)

Leo: Or can Stan … call upon other sources of information?

SR: Robert's cabin?

Leo: Robert's cabin … yes.

SR: Yeah … I remember … the dogs were barking … I remember the dogs were barking … dogs … dogs were barking. And, um … I woke up—it was morning. I think it was four or five. I went to see what the dogs were barking at. Robert's dog was downstairs. I went downstairs and saw that the door was open. I thought Robert had opened the door. And the dog was keeping a distance and … I remember Heidi asking me what was going on. I said I didn't know, that the dogs were barking at … something outside. She was behind me on the steps and … I looked out the door—there was something standing outside the door … off to the right a little bit … by the wooden pole, the wooden post. And … it wasn't Robert; it was small. It was small. It looked like … one of the little guys that was … that I'd seen before [unintelligible]. And the dog started … going toward this thing, growling, and it had a silver … silver cylinder. And ... I heard a "crack" come out of it like a small ... firecracker. And I saw an energy—a massive energy—go whizzing right by my leg. I could feel a tingling in my leg. It hit the dog, and ... the dog rolled around just yelping really, really loud, rolled around and ran up the stairs and ... went to the bathroom ... on the carpet, ran up the stairs and ... I could hear Heidi going, "Oh, my God!" And ... I don't remember after that. I don't remember anything else after that. I remember ... I think we were ... Heidi was with me. I think we were ... in a room somewhere, but I don't remember much. Just ... and there was another person there, but I don't remember much. A kind of weird pink light, kinda-sorta, I've seen the pink light before, but ... I don't remember much.

Leo: OK. What do you remember next? What happened next?

Grandpa: *(Long pause)* It is not important.

Leo: Not important?

Grandpa: It is not important.

Leo: Are we speaking to Grandpa?

Grandpa: Hello, Leo.

Leo: Good to have you here. Thank you. Do you have information that you can share with us through Stan about what happened on August 11?

Grandpa: It is not important.

Leo: It is not important? Is there other information that is important for us to have?

Grandpa: Concerning August 11?

Leo: Yes—or other matters?

Grandpa: Yes. There are other matters that need to be addressed, but August 11 is not—it is not important.

Leo: Well, we are thankful that you're here, and we would be glad to receive information from you. What matters are more important?

Grandpa: You have questions.

Leo: Yes, we do. M-hmm. Are you willing to accept questions [and] respond to them?

Grandpa: If allowed, yes.

Leo: All right. I have some written questions from Susan Carson.

Grandpa: Yes.

Leo: One question she asks … or … it's not a question; it is a request. And she requests that you describe in detail the area where Susan keeps material related to Stan's case. Do you have that information?

Grandpa: Hmm … we'll see. (*Pause*) She is no longer part of this. But … some she keeps in … some … most has been returned to Stan. There is not much left that Susan keeps.

Leo: I see. OK. She has another comment and request. She said there was an alteration to Susan's house which suggests that somebody had been in her home on the day she discovered pieces were missing.

Grandpa: Stan's implant.

Leo: Oh, is that the alteration? Do you know the—

Grandpa: Stan's implant is missing.

Leo: Do you know the container which held the implant?

Grandpa: It was a glass jar.

Leo: Do you have other information about what happened?

Grandpa: No.

Leo: Do you know—the glass jar—do you know the size … of the jar?

Grandpa: Hmm … (*gesturing with hands*) … about … *word* … inches … inches … approximately three and a half inches by … maybe four inches. Three inches tall by … only two inches—top narrows toward the bottom.

Leo: OK. Any other information …

Grandpa: No.

Leo: … about what happened?

Grandpa: No. We do not monitor everything that happens.

Leo: OK.

Grandpa: It is not necessary.

Leo: All right. Thank you. And … some questions from Alejandro. He's asking, "Is there an equation that you can share at this time?"

Grandpa: It is not necessary.

Leo: Not necessary at this time? OK. Are there others who are using quote "Audrey's" voice, making calls?

Grandpa: Yes.

Leo: Do you know who and why they are making these calls?

Grandpa: Hmm ... to disinform, to confuse ... to confuse Stan and others involved. But Stan knows, everybody knows. They must trust their instincts to ... guide them.

Leo: So these probably are persons who are attempting to confuse or to disinform?

Grandpa: Yes. It has happened. Not often, but it has happened. Most are ... communication. Most are ... real and some are not—not many, though ... are not.

Leo: Is there further information you can share now about the true messages from the voice, Audrey?

Grandpa: Hmm ... warnings and ... new news.

Leo: Communication to help Stan in his work?

Grandpa: Yes ... understandings and ... *word* ... to help guide.

Leo: OK. Thank you. And is it OK to respond to questions from people here?

Grandpa: Yes.

Leo: Are there some questions that people care to ask Grandpa now?

H: Well, I'll start with one. I would like to know why it is that you are not allowed to meet with our group face to face. I understand why you are not allowed to show yourselves to everybody at this point, but not only would we very much appreciate the experience, but that would certainly solidify a lot of things in people's minds, and I'm just curious as to why it is you're not allowed to meet with us—face to face—yet.

Grandpa: Hmm ... we are not allowed because you are not ready. We must make sure that ... there is no fear. We must make sure that ... humans are ... still a young race. Humans are primitive and when there is fear, there is aggression—sometimes aggression that cannot be helped ... and ... there still is fear and we can sense that.

H: Within this group?

Grandpa: Yes. And some people not so much, and some more so. Even Stan has some fear. But when it is time ... when it is time.

Leo: Want to quantify that? Do we have an assessment of when that time will be?

H: Exactly! I mean, are we taking any time in the—

Grandpa: It is not up to us; it is up to you! When people are more accepting—humans do not even accept themselves! It is more complicated. Humans don't quite understand it … it will be sooner than later.

H: Well, do you feel that it will be within the next, say, six months?

Grandpa: It is up to humans—it is up to you. It is not up to us. You will know when it is right; you will know when it is time!

Robert Morgan (RM): What can we do as human beings to come closer to accepting you?

Grandpa: You are on the right track. You are doing what you are supposed to do. Hmm … when it is no longer about greed, when it is about caring and understanding, then you will know. When it is no longer about fear, but when it is about enlightening—then we will know. There are those even here that … are close, but still have … somewhat their own agenda. That will change, we know, but … it is not yet. It will be, though.

Leo: Alejandro, do you have questions?

AR: The Nebraska video, another video of an entity—

Grandpa: (Giggles)

AR: Why do you giggle?

Leo: (Laughter)

AR: Why do you laugh when we refer to that one?

Grandpa: It was amusing!

AR: How so?

Grandpa: Stan is funny. (Laughing)

AR: What about Stan amuses you?

Grandpa: His reactions.

AR: Is the video then truly one of your people?

Grandpa: Yes. Hmm … one of our people, one of our people—not exactly of our people, yes.

AR: But an extraterrestrial—

Grandpa: What you call extraterrestrial … extraterrestrial …

AR: Is there a better phrase than "extraterrestrial?"

Grandpa: We are all connected. Everything is connected. We are all part of oneness. Just … "extraterrestrial" is fine.

AR: Do you know why that entity was there?

Grandpa: To observe, to monitor Stan's reaction. It was not by accident.

AR: We've been asked that question quite frequently.

Grandpa: We understand—we know. People that do not believe, that is their right, but ... it is to—Stan knows this—it is to start the conversation. It is to start the spark. And then we'll go from there.

AR: Are you aware that video got on the Internet?

Grandpa: Yes! We know!

AR: Do you know *how* that got on the Internet?

Grandpa: Hmm ... yes. We ... are almost ... *word* ... *word* ... we have an idea. But are we positive—a hundred percent positive? No, but we have an idea. Stan is on the right track. But we are not sure if it is from ... *word* ... from ... *word* ... it is given to them to put on the ... (slowly) *Inter- Inter- Internet*.

AR: The original monitoring call which we received alluded to being a government insider.

Grandpa: Could be. Government isn't what you think. Government is ... there are ... many ... *word* ... (slowly) *factions* within ... government. One does not know what the other one is doing. In most cases there is ... so much corruption. There is so much ... misguidance that ... it will change soon. It will collapse. A house that stands ... stands ... stands ... divided.

AR: Thank you. Do you all now use that Audrey—what we call "Audrey" technology?

Grandpa: It is hard to communicate. We do not communicate like humans. Not always, but ... not ... us personally, but ... we are ... we are a council ... council. You know about this, we have talked about this before.

RM: Now, if you won't meet with us personally, would you allow us to photograph you again?

Grandpa: If needed. And I did not say we would not meet with you personally. I said it is not time yet.

RM: Because if we could get a picture of you walking, then—

Grandpa: I understand your motive, yes. It's not needed right now. Maybe soon ... when it's needed. You are on the right track, but motives have to change somewhat. I understand, but ... it has to be about enlightenment. It has to be about ... egos must go away.

RM: But if you would give us that picture it would help us enlighten humankind, and then they would know that you are real—

Grandpa: If we decide it is needed, it will happen. If it's not needed, it will not happen.

H: Wouldn't it help greatly to do what Robert is asking—to give us another picture …

Grandpa: No, it is not necessary.

H: … that cannot be questioned?

Grandpa: It's only for … it's only for your insecurities. You want proof. You are on the right track!

H: Well, we know that but others don't—those whom we are trying to enlighten.

Grandpa: We are not allowed at this point. Humans are not ready for … what you ask. We can … periodically, (slowly) *periodically*, periodically … do things to assist, but … humans want sovereignty. They have to come to this themselves. Humans want … *word* … to run themselves … to run, to run, to … they want *sovereignty*. They must show their worth. Many civilizations have come to this point. Many have reached past this point; many have failed and perished. [The] human race is at your … cross … (*gesturing, crossing fingers*) …

RM: Crossroads?

Grandpa: Crossroads! Thank you. Crossroads … and … it can go either way. We are here to guide as much as we can, but we cannot get directly involved. You know this; we've talked about this.

RM: But through Stan's story, it's going to help us enlighten human beings.

Grandpa: It is already happening. Everything has been guided. You know this also. You ask for stuff constantly, but … you must understand, it must be subtle. It cannot be … Stan also expects … *word* … to land on the lawn … to land on house lawn …

Leo: The White House lawn.

Grandpa: … White House lawn. It will not happen; it has to be subtle. Humans are not ready. They are getting ready, and when the shift happens, everybody will know.

RM: What can I do to help?

Grandpa: Continue supporting Stan and Lisa. Understand and … (slowly) *understand* this is about enlightenment. This is not anything but enlightenment. If it is about … self-fulfillment, it will not happen. Everyone involved in this must understand this is for everyone. This is for the good and the bad. It is for the people that want to stop it, as well as the people who want to help. This is for everyone. It is about ... Stan ... Stan's words ... good word ... global, global ...

SD: Consciousness.

Grandpa: Consciousness. Global consciousness. As that change[s], everything will change. It will be ... unexplainable.

H: Do you see great strides or—

Grandpa: Humans do not understand who they really are. Humans are more than they know. We are still studying this. Humans are greater. Humans are a water vessel. Humans are a water vessel and more. They are a ... humans are ... a physical expression of ... *word* ... deity ... deity ...

H: The Divine?

RM: Celestial beings?

Grandpa: Divine. Divine. Humans ... humans are ... humans are more than they know.

RM: How can we find out about what we really are?

Grandpa: Experience ... just experience.

H: Do you see great strides being made toward the shift? For example, since the last regression? Have you seen—

Grandpa: It is all ... planned, everything that has happened. Stan knows this.

H: I'm talking about the entire population.

Grandpa: Yes! It is happening *now*!

H: I know that.

Grandpa: Then why are you asking?

H: What I meant ... from your point of view ...

Grandpa: Yes?

H: From your—

Grandpa: We see it, yes. More work needs to be done, but we hope that it will happen sooner than later. It is ... getting close. It is ... cutting ... *word* ... cutting close ... (*gesturing*)

Leo: Cutting edge?

Grandpa: It is very close.

RN: You had told us a year or two ago that we were about thirty percent, as far as the number of people—

Grandpa: It is more now.

RN: Do you have any idea of the number?

Grandpa: Hmm … no, it is more, but there's still a lot more … a lot, lot, lot … there is much more to do. It is … is it getting closer. We hope that it will happen but … we are cutting it … cutting it close …*word* … is cutting it close.

RN: Thank you. Now the recent implant in Stan—did your group place that?

Grandpa: It is not important.

AR: Do you know how Stan can help his health situation?

Grandpa: Hmm … we are still studying that. We are confused. We thought it … we thought it was from us, but we do not think so. It might be something else, genetic or … something government or … Stan is being monitored. He has … (*gesturing toward my abdomen*) …

AR: Stomach?

RM: Internal?

PW: Immune?

Grandpa: … um … off … (slowly) *endocrine, endocrine* … endocrine. We are not certain it is from us now. We assume that it might be, but it doesn't appear to be; it might be something different. It might be genetic; it might be … we cannot get involved, though. We feel Stan will have this figured out, but Stan is afraid he will … (*smiling, beginning to laugh*) … he is afraid.

RN: Are there any other entities that either operate or function through Stan?

Grandpa: It is not allowed.

H: The information—or it's not allowed for it to happen?

Grandpa: Not allowed. It is … Stan is … closely monitored.

H: So you mean it's not allowed for other entities to work through him? Or do you mean it's not allowed for you to tell us that?

Grandpa: It is not allowed for other … what you call "entities" to … work through him.

Attendee: I have a question. Many on the planet are predicting financial collapse and economic chaos.

Grandpa: Yes!

Attendee: Is that …

Grandpa: Yes!

Attendee: ... a necessary part of the process?

Grandpa: Yes! No, it is not us; it is government. They are ... trying to control, trying to control populations. They are ... they have a plan. This is what we are talking about. Time is short. They are ... these are evil ... *word* ... *good word* ... ... groups ... no ...

PW: Cabal?

Grandpa: Yes! Yes. Thank you, Paul.

Attendee: Another question on that: Is there a possibility that this will occur very soon? Is there ...

Grandpa: We do not know

Attendee: ... planning?

Grandpa: We do not know. It could happen sooner; it could happen later. We do not know. We know that there are plans. It will start with road ... road ... road ...

H: The highway?

Grandpa: Highway, yes. Road ... road ...

H: From Mexico to Canada?

Grandpa: Yes, yes, yes, yes. From pole to pole.

Attendee: In the timeline of this planet, how long have you been observing us, and how many civilizations or entire civilizations on the planet—

Grandpa: Hmm ... us personally ... we have observed for a *very* long time. We ... different groups, not ... us but ... different groups. We have watched rise and fall, rise and fall ... (*removes my digital camera from my pocket and takes several close-up pictures of self*).

Attendee: Do you see any cycles [unintelligible]?

Grandpa: Yes.

Leo: How many star gates or portals are there?

Grandpa: It is not important.

AR: Why do you like to take pictures of yourself?

Grandpa: Pictures ... pictures ... pictures?

AR: Yes. We've gotten other pictures of entities, entities that take pictures of themselves.

Grandpa: Hmm ...

RN: You can take a picture of us if you like. You can show them to your family!

Grandpa: (*Softly*) Pictures? (*Tosses camera onto floor*)

Group: (*Gasps*)

Leo: (*Laughter*)

Grandpa: Pictures ...

Victoria (V): Recently there were some threats made on Stan's life, an attempt made on Stan's life.

Grandpa: Yes. We know.

V: Were those perpetrated by the government?

Grandpa: Hmm ...

V: By *our* government?

Grandpa: Hmm ...

V: Military?

Grandpa: Hmm ... govern—don't understand ... you don't understand government, but, yes. Military ... they are afraid, but ... Stan is doing the right thing. When Stan is able to ... get funding, when Stan is able to ... get this out, you will no longer have to worry. Getting the book out is good. It is a good thing because they ... try to stop it. And it ... it is too late. It will be too late. When ... *word* ... *word* ... when ... it gets out, it will be too late. We are ... we cannot get involved. It will just happen soon, for Stan's safety and for Lisa's safety and for everyone. It's not just Stan now. They are interested in everybody now because they are afraid, but they know that their time is short. They are panicking. We are monitoring this. Stan must not go on walks anymore. That was foolish.

V: When he did take that walk, how is it that you know that he thought about walking to Denver?

Grandpa: It is not important. (*Long pause*) You know.

Attendee: Do you know that I am writing Stan's story for him?

Grandpa: Yes.

Attendee: Am I on the right track with it?

Grandpa: Hmm ... yes. It would be ... *word* ... prudent to make sure that ... Stan's words are not lost. Do you understand?

Attendee: Yes.

RN: What are some of the largest misperceptions humans have about who they are?

Grandpa: They feel they have no control, when they do. The must learn to trust their instincts. Their instincts are ... valid (*momentarily holding one finger up as if to pause questions*). OK.

RN: What abilities do we have that we are not currently using?

Grandpa: Communication. Humans' communication is slow. There is ... we have discussed this before. Humans have abilities they do not understand. There are those in this room that have abilities that are greater than some. But it is turned off. It is turned off. We are studying this. We still do not understand, but ... when the shift happens ... we are sure it will all change.

Attendee: What do you mean by the shift?

Grandpa: Shift, shift, shift ...

Attendee: In consciousness.

Grandpa: Yes—consciousness. Thank you.

Attendee: The consciousness shift that will happen at a certain time?

Grandpa: Yes. Hmm ... global—Stan's words—global consciousness. People do not understand. Humans do not understand how connected everything is. Time is not what they perceive. Time is different. Time is now. Time—past, present, future—are all one! It is like a loop. It is like a ... like a loop. Everything is connected. We are all connected. We are all part of the oneness, a oneness that is fractured.

CS: This is Claude. Is there anything I need to know from you right now?

Grandpa: Continue to eat.

Leo: (Laughter)

CS: Is it important for me to get the theories finished?

Grandpa: You are on the right track. It is about the learning. It is about supporting Stan. You must know that there are going to be ... when it's ... not by accident, it's by design of the people that do not want this out. You're about to be ridiculed. You have already been ridiculed, but it will get worse—but you must stand firm in what you believe and do not let them get to you. Stan goes through this all the time. He is going through this now—Lisa and Stan and the family. This is part of it, but this is also ... you must understand that this is also ... *word* ... bringing attention and bringing attention is good.

But do not let them get you down. You are on the right track. Stan ... has ...

CS: Do you know the work of Dolores Cannon?

Grandpa: Hmm ...

CS: She's a hypnotherapist and talks about the possibility of a splitting of reality into separate worlds ...

Grandpa: There are many ... many different realities ... in all the same place. Humans don't understand. She is ... I do not ... we do not know of her, but ... I understand the concept. It has been talked about before among humans. There are many realities that occupy the same space. There are many, what you call "dimensions" that occupy the same space. There are many things living in the same space, just in different vibe—what you call "vibrations." Vibe ... vibrations, vibrations. Humans don't quite understand.

RS: I have a question. Everyone in this room has a physical role in helping Stan.

Grandpa: Yes.

RS: What is mine?

Grandpa: To support Stan. You role has not come to fruition, fruition ... but you are an important part of this. You helped Stan and Lisa already with ... their house, their house, their house. It is not by accident. Everyone here has been chosen for a reason, everyone. There are people that have been involved with Stan's life longer than Stan even knows. And there are people that ... have not yet come into Stan's life that are a part of it, but ... it is all part of a grand ...

RS: Scheme? Program?

Grandpa: ... design.

RS: On the fear aspect, how acute, how sensitive are you to feeling this fear? What do you get from feeling this?

Grandpa: Fear is a primitive emotion. All beings have fear, but it is how it is controlled. Lisa has fears; Stan has fears. Their fears are unfounded. Everybody has fears.

RS: But how sensitive are you to ... how do you react to that fear? How sensitive are you to it?

Grandpa: Human fear is painful because it is primitive. They do not know how to control. That is why we do not ... *word* ... suddenly appear.

RS: Right.

Grandpa: As we have discussed. It is because of humans, not because we do not want to.

RS: I understand.

Grandpa: It is a matter of trust.

RS: Thank you.

V: Did you make an appearance at this last party?

Grandpa: Yes.

V: Twice?

Grandpa: More than twice. When you were sleeping.

V: Is it a good idea to set up speaking engagements for Stan around the country?

Grandpa: Yes, it is to be.

V: To share this information?

Grandpa: Yes.

V: Well, he needs someone who may invest in his documentary, through the speaking engagements.

Grandpa: We believe somebody is already here but, yes, that is a possibility.

V: Are you familiar with the upcoming Democratic National Convention?

Grandpa: Yes.

V: A lot of people from all over the world will be here. Is there a chance that you can make an appearance?

Grandpa: No, it is not necessary.

V: Is Denver the location for the United States shadow government?

Grandpa: There are many shadow governments. Denver is just one location. It is not just this area. They are all over the world. They are ... *word* ... Paul!

PW: Yes, sir.

Grandpa: Word. There are ... all over the world, groups ... not groups ...

V: Factions?

Grandpa: Factions! Thank you.

PW: Grandpa, do you know if the loss of my last job ...

Grandpa: Yes.

PW: ... is related to Stan ...

Grandpa: Yes.

PW: ... my connection to Stan?

Grandpa: Yes.

PW: What happened?

Grandpa: You lost your job.

PW: Is this connected to my relationship with Stan?

Grandpa: Hmm ... you have more important things to do.

PW: Good. Thank you. Can I be more helpful to Stan?

Grandpa: Yes, you are. There will be more books. You will be involved. You and Stan discussed this. You are helping Stan. It's your support that's helping Stan.

PW: What's your assessment in my control of a paranoia factor, as we discussed previously?

Grandpa: You are learning, but you have much to learn still.

Group: (*Laughter*)

PW: Thanks.

RN: Hey, Grandpa, is there a special practice that we could use to attain enlightenment in a shorter space of time, or do we generally need to wait for the shift?

Grandpa: In this room, you are on the right track. Everybody here has been chosen for a reason. Everyone here must know that it is about the enlightenment. It is about caring. It is about the oneness. Everyone here must know that it is ... that everybody is connected. That your actions affect the next person, affects the next person, affects the next person. It is about ... even when you do not ... *word* ... even when ... *word* ... you must ... this is for everybody. This is for the people that do not want this. This for people that are against Stan. This for everybody. Stan knows this. And ... everybody in this group must know and must support each other. You must support Stan and Lisa, and Stan and Lisa in turn will support everyone in the group. That is part of ... the process. That is part of the enlightenment and ... Stan's words ... you must ... example by ... setting example by ... giving it by ... *word* ...

V: Lead by example?

Grandpa: Lead by example! Thank you!

RS: Grandpa, I have a question. Are you familiar with the book that I am writing?

Grandpa: We have heard you talk about this, yes.

RS: And will this help in the shift process?

Grandpa: I do not know.

RM: You talk about enlightenment, and you talk about fear in us? Well, we fear what we don't know.

Grandpa: We understand that.

RM: So why can't we get to know you better ... so we won't fear you? Why won't you help us do that?

Grandpa: We are not allowed. We cannot get involved. We cannot ... humans want sovereignty. Humans must grow up on their own.

RM: How can we help ourselves "grow up" faster?

Grandpa: You are on the right track. There is no "fast." It will happen when it is time. We hope that ... you are at the stage now where humans ... can go either way ... Stan's words ... either way, (slowly) *either way*. It will happen, either the right way or the wrong way. We hope it will happen the right way. There are many benefits to all if it happens the right way.

Attendee: You say you cannot get involved, but when you abduct people aren't you involving yourself?

Grandpa: Hmm ... abduction, abduction ... not a good word, "abduction." We try not to ... there are reasons; humans wouldn't understand. Humans would not understand. It is not just for humans that we do this. It is for ... there are some that have their own agenda. There are some ... like us that do not ... that are concerned about what is happening. We try to make little impact. That's why most do not remember ... some that do. There are those that have been ... have experienced their own kind to ... *word* ... disinform ... to scare away. It is frightening for some, yes. Stan is frightened but ... there is a reason.

H: Grandpa, some time ago I had a call from Audrey that indicated I have offspring off-world.

Grandpa: Yes. Yes.

H: How many do I have?

Grandpa: One. Do you want more?

Group: (*Laughter*)

H: I thought you had information. Whatever needs to be!

RN: Are humans an experiment?

Grandpa: We do not know. Um … they are more. They are a conglom-eration. They are … *word* … we are studying this.

RN: Have you discovered anything new about humans or—

Grandpa: We are discovering new things every day.

RN: Could you share any of those?

Grandpa: No. Humans are bizarre sometimes, but it is because they are primitive. Stan is … still questioning when there is so much evidence. That confuses us. Stan and Lisa have fears about each other, but they're unfounded. That is confusing. We are learning.

AR: Are there secret human organizations …

Grandpa: Yes.

AR: … battling evil Greys?

Grandpa: Greys … Greys …

AR: Entities.

Grandpa: Hmm … evil … confused … we are confused.

AR: You used the term "evil" earlier, referring to humans.

Grandpa: Yes, but … agendas—it is about agendas. Just because a meteor crashes into a planet doesn't mean "evil" had guided it there. It just happens. There is … there is good and bad out there, as there is here, but it is not the same. You have to—Stan knows this—you have to … you have to … in order to reach … a more advanced level, it is … *word* … a test. Stan knows this. Most everyone here knows this. There are those that do not want this enlightenment of mankind, and there are those that do. Just because there are those that do not, does not mean they are evil. It just means their agenda is different.

RN: We are also studying your race and many others, trying to better understand. Is there anything you would be permitted to share with us about your race that we might not be clear on?

Grandpa: We are studying *your* race … and others.

PW: Grandpa, this may sound like a crazy question …

Grandpa: (*Laughs*)

PW: … but are there races like the dragons or reptilians in appearance that are … that are drinking human blood?

Grandpa: We have discussed this!

PW: Then … we determined that I was correct.

184

Grandpa: There are many races, and there are some that resemble this, yes.

H: But are they here, living in the—

Grandpa: There are many races living here. Humans have no idea!

Matt: How do different races survive in our atmosphere with our bacteria and things of that nature?

Grandpa: There are some that have been here a very long time. When I say many ... maybe ... there is seven or more, not ... hundreds ... but there have been colonies here—what you call "colonies"—here for as long as man ... and even before man.

RN: Are there races under the surface of the Earth?

Grandpa: Yes. Not how you think. Not ... not ... how you think.

RS: How do they live under the Earth?

Grandpa: They are ... dwellings built for them.

RS: Tunnels?

Grandpa: It is not important.

Matt: Did we acquire some of your technology?

Grandpa: Yes.

Matt: By shooting down or obtaining that from crashed vehicles?

Grandpa: Not you, but government, yes. There have been situations like this, yes.

Matt: Did you provide some technology?

Grandpa: Not us, but there are some that have. And there are some that have their own agenda that ... are not on the right path that have agreed, yes. This will soon change. This will ... all change.

RN: Could you identify some of the technologies that were derived from ...

Grandpa: No.

RN: ... Roswell crash?

Grandpa: No, it's not allowed. It's ... it's ... it's ... Stan's words ... it's ...

AR: Lisa has some questions. She doesn't want to ask them, though.

Grandpa: She must!

AR: Here are her questions.

Grandpa: Lisa.

LR: (No response.)

AR: [Lisa's daughter] had some concerns about our scientists experimenting with the super collider and testing dark matter. Are you aware of this?

Grandpa: Yes!

AR: Do you know what the possible outcome will be?

Grandpa: Yes. It's not good!

AR: If it goes wrong will you intervene? Will you intervene if—

Grandpa: We have not decided yet. But this is ... not a good thing. Humans are not ready. They are not ... ready. They are ... *word* ... playing ... playing ...

H: Playing with matches?

Grandpa: With fire, fire. Playing with fire. Matches... matches ...

AR: The caller we call "Audrey," can you tell us who he or she is?

Grandpa: It is not important.

Leo: Why was that voice chosen?

Grandpa: It was chosen by others before. We find it easy to manipulate and use. Humans do not communicate like we do. We have to slow down. We have to find something to communicate ... so we manipulate. It is easy to manipulate. And Stan is familiar. We borrow ... (slowly) *borrow* ... *borrowed* ... the voice from the first call because Stan is familiar.

H: Why is it easier to make phone contact certain times of the year versus others?

Grandpa: It is a matter of ... alignment. Humans would not understand. It is matter of alignment. Great distances, it is hard to communicate.

H: Does it have to do with the appearance of the Orion constellation?

Grandpa: Hmm ... more than that, but ... close. Very good!

RN: Are you going to have to be away for a long time again?

Grandpa: Hmm ...

RN: In the near future?

Grandpa: Yes ... we will be here a little bit longer this time, but not like last time.

AR: Lisa has quite a few more questions.

Grandpa: Lisa!

Group: (*Nervous laughter*)

Grandpa: (*To LR*) Ask them.

186

AR: Stan asked her recently—

Grandpa: No! No. Lisa.

LR: (*Refusing*) I'm not going to ask them then.

AR: Are you alluding to the first Audrey call not being you, then?

Grandpa: No, it wasn't us. It was ... it was ... good, good, good ... good.

RN: Are the calls that were not you, are they ... do they have Stan's best interest and human's best interest in mind?

Grandpa: First ones, yes. There have been two that were meant to help, but they were not us. There have been some that were to confuse.

Leo: I think I may know the answer, but as [we] receive these calls, is there any way that we might know whether they are—

Grandpa: Most are ... most are real. Some are not. There have been a few of ... the people here who have gotten a fake sampler, fake call.

RN: Can you identify any of those calls that were not yours?

Grandpa: It is not important.

Leo: [Unintelligible] the fake calls?

Grandpa: No.

H: No?

AR: Why did you remove the phone from Stan's house?

Grandpa: Monitoring ... monitoring device. We monitor to help Stan eliminate this. We use something easier for them.

Attendee: Are there any other frequencies we need to investigate to prove your existence to the general public?

Grandpa: No. You are on the right track, it will happen. It must be in steps. (Slowly) *baby ... baby* steps. It cannot happen all at once. There are those that are not prepared. This is for everyone. Mostly out of fear. People are guided incorrectly because of fear. Fear is guiding the world. There are ... few that rule the many. This must change. It is all about fear. They will try what they can, fake ... fake—Stan's word— fake ... what they can to ... create fear, to control. But ... people instinctively know that this is not the right way. They must learn to trust their instincts.

RN: Long ago you had told Stan that there were seven.

Grandpa: Yes. There are seven.

RN: What is the status of that? Have you—

Grandpa: It is changing. As we said before, some will come, and some will go. The seven ... there are still seven. Some ... we thought would be part of it are no longer part of it. Some ... have not come into play yet.

RN: How can we find the seven?

Grandpa: Hmm ... it will be—Ben, Stan ... there are others ... Robert ... there are others.

PW: Victoria? Is she one of the seven?

Grandpa: Hmm ... we do not know.

RN: What is the purpose of the seven?

Grandpa: To help ... to help enlighten, to help guide. There are ... there are ... they have different ... *word* ... they have different ...

Leo: Paths?

Grandpa: Yes! That ... help ... strengthen ... help strengthen each other, help strengthen ...

RN: Contribute?

Grandpa: Contribute to each other—that will make ... like pieces of a puzzle.

RN: Can you share the components that are missing right now?

Grandpa: No.

AR: When you removed the phone that was monitoring ... why did you need to remove the phone if all the information is for everyone?

Grandpa: Hmm ... there are certain people ... certain groups that ... will not be enlightened. That no matter what happens ... *word* ... that are concerned about their own ... egos and their own ... they are losing ground; they are panicking. They will never be enlightened. We want to protect Stan and everyone involved from these people.

AR: You said in past visits that Stan is a role model. You also said that you have the ability to control humans to some extent. If this is true, are you responsible for Stan's angry outbursts with people, and why can't you help those?

Grandpa: No. Stan is frustrated. He is confused. He tries, but ... he is not ... *word* ... although Stan ... misunderstand ... Stan ... *mystery word* ... *mystery word* ... angry, frustrated ... he is ... stressed. This is a great responsibility. We try not to overwhelm, but sometimes it happens. Stan is overwhelmed just like ... the people in ... Stan's family are overwhelmed. This is unfortunate, but a by-product. It will pass if they stay strong. There are unfounded fears. There are fears that should not be there. They need to go away. It creates anger and frustration and ... that is a vicious cycle. It must go away.

AR: Stan has children that he's seen once, living among your people, and he worries about them. Can you tell him anything that will help him not to worry about them?

Grandpa: They cannot survive with Stan. He knows this. They are well taken care of. What we are learning from Stan and others helps us to care for the offspring. We are trying to learn ... that is the process. We are learning, as Stan and others are learning from us and ... and we hope that sometimes here and there will be a ... agreement to exchange and learn from each other.

H: Audrey indicated that Victoria also has children.

Grandpa: Yes.

H: How many?

Grandpa: As many as Stan.

H: Are they hybrids?

Grandpa: Yes.

H: All of them?

Grandpa: Yes.

Leo: Are all the children there hybrid—

Grandpa: Yes.

Leo: I mean, are any one hundred percent human?

Grandpa: No.

H: And will we be allowed to meet them, Victoria and I?

Grandpa: Sometime, yes. It is not what you expect. They do not think like humans think. Humans need ... nurturing ... nurturing ... need nurturing in relating. At first we were confused, and there were some that had perished because we did not understand, but ... not like humans. These would not survive. These are ... they are accelerated. They are ... they are ... children need ... human children need years. These ... surpass that.

RN: Are some of your hybrids here on Earth now?

Grandpa: It is not ... it is not important.

189

SD: Do you guys do this in order to acquire some of our abilities, some of our traits as humans by mixing them with your type?

Grandpa: It is not important.

RN: What genetic characteristics do humans have that interest you?

Grandpa: They are a sturdy race.

RN: Physically sturdy?

Grandpa: They heal fast; they are a studier race.

AR: [Stan's stepson] is curious about the writing on his wall that was his experience. Can you tell us how this was accomplished and why it happened to an eight-year-old?

Grandpa: Stan … doubts … doubts … Stan doubted. Stan needed to know that this was real. We chose this to prove to Stan that was no doubt. [Stan's stepson] was unharmed. [He] … is being enlightened. It is all part of the process.

AR: [He] also wants to know, a couple weeks ago, you were visiting their home when they were hearing noises like thunder echoing through the house—

Grandpa: We visit Stan often.

AR: You make the sound of running on the roof?

Grandpa: (*Grinning*) Hmm. It is not important.

H: That's important.

Grandpa: There is … evidence, just recently …

H: At Stan's house?

Grandpa: Yes.

H: Did you scatter the kitty toys around the living room?

Grandpa: (*Giggling*)

H: I guess that's "yes."

Grandpa: It is not important.

SD: Why did you not give Susan Carson the crop circle that you promised?

Grandpa: She is not part of it. She chose to leave. She is no longer a part of it.

SD: Is the information in these regressions limited personally for our [unintelligible]?

Grandpa: That is correct.

SD: Can I just feel your hand or something to see like … if it feels strange?

Grandpa: (*Extends his hand*)

SD: (*Laughing, approaches me and touches my hand*)

Group: (*Laughter*)

SD: (*Laughing*) You feel normal.

Group: (*Laughter*)

SD: Is there any way that we can do something here today to sort of confirm some of the information here—we are even doing that, sort of, I guess with Susan Carson's question, but is there something that maybe you can … do for somebody in the room or … no?

Grandpa: (*Grimacing*) Hmm … you just want us to perform. We do not perform, we do this to …

SD: I need some more credibility.

Grandpa: You are on the right track. [Unintelligible] you must stop questioning yourself. You know inside … *word* …

SD: OK, well, I'm not doing it just for myself. I'm trying to convince others.

Grandpa: You understand more than most. You have gifts. You know about …

SD: OK. I'm writing a book, and I'm trying to explain some things about space—and about the creation and expansion of space. And I was wondering if you were familiar with any of these theories, and if they're even slightly—

Grandpa: Space is not empty.

SD: Hmm … OK. Is the expansion of the universe primarily due to consciousness [unintelligible] in the universe?

Grandpa: You would not understand.

SD: OK. Can one dimension expand another, from one another's influence?

Grandpa: Dimensions … dimensions … what you call "dimensions" interact with each other. That is why … nuclear … nuclear … nuclear bomb … nuclear atom … bomb … atom bomb … they affect more than just this dimension. That's why we are so concerned.

RN: Can we as humans, individually, affect other dimensions?

Grandpa: Humans … humans … humans … it is more a global consciousness, but yes … Stan's words … butter … butter … fly …

RM: Butterfly effect.

191

Grandpa: Butterfly effect. That is ... that's what it is hard to ... past is easy—future is hard. Past is easy; future ... the future is constantly changing.

SD: But I'm seeing the universe as just this sort of large vacuum and exchange of energy—one big constant change.

Grandpa: It *is* one big constant change. Yes, you are very smart. But space is not empty. Space is full.

SD: Did you fly over my house?

Grandpa: (*Giggles*) *And* Stan's.

SD: Why did you do that?

Grandpa: To ... to ... further ... qualify Stan's experiences.

SD: OK. Thank you. What does "enlightenment" mean to you?

Grandpa: What does enlightenment mean to *you*?

SD: Well ... correct me if I'm wrong, but I'm discovering super-human consciousness that I assume would be called, "enlightenment."

Grandpa: That is part of it. It is not "super-human"—it is natural. Humans don't understand who they are.

SD: But on reaching this consciousness, there are [unintelligible] us to develop and get hormones for the human body?

Grandpa: That is ... human attempt to understand the deity ... the deity ... the deity, the oneness. It is more than that. That is just a physical aspect of ... the reality here. There is more.

SD: OK ... because they say that reaching this actually slows the process of aging and it [unintelligible] throughout the human body.

Grandpa: Aging—that is ... that is physical. That is ... physical—human. There is no ... aging. There is just change.

SD: So do you age, as humans age?

Grandpa: Yes ... not like humans ... longer.

SD: Does your race have disease?

Grandpa: Yes. All physical races have disease. We are physical. We are more advanced, but we are physical. We understand the non-physical. We understand the ... *word* ... it is not important.

RN: Are we more spiritually evolved than your race?

Grandpa: Humans ... humans are ... we are studying this. Stan is tired; he needs a break. Sugar ... sugar ... sugar ...

Leo: OK. Thank you very much for responding to questions. And thank Stan for sharing that information through him. And we will talk to you later. Thank you very much.

After a break, the session resumes.

Leo: Greetings, Stan Romanek—Leo Sprinkle. It's almost four-thirty, August twenty-third, 2008. We'll ask Stan to relax deeply, and we'll ask Grandpa to speak through him.
Grandpa: Hello.
Leo: Greetings. Thank you for being here. Are you willing to accept other questions from—
Grandpa: Yes!
Leo: OK. Thank you very much. Another question has to do with information about Dr. Steve Greer, the physician who implies or has said that the government "cabal" or the shadow government has created beings to look like you. Is that true?
Grandpa: Not ... yes, they have technology that is higher than most know. They are not very good at it, but ... yes, they have tried.
Leo: Do you know the purpose of that?
Grandpa: Yes, to ... scare, to ... disinform ... to—we are aware. It has not happened often, though.
Leo: I have a question for you. In an earlier session you had described your group as being at a third or half level between one and one hundred intelligence—humans being at one, and various extraterrestrial or extra-dimensional groups being at one hundred. If I recall correctly, you said that your group would be at 30, 40, 50— something like that. Does that represent the level of intelligence that you perceive as your group?
Grandpa: It is not just intelligence, it is ... evolution. The more evolved one is ... the less physical ... are. There is ... beyond us ... we do not understand. But ... we are forty percent.
Leo: Forty percent?
Grandpa: Yes.
Leo: A related question: Do you work closely, or do you not work with the other groups at various levels?

Grandpa: We do at ... some slightly ... varied levels, but not ... we do not understand the higher levels. We are studying. We seem to be guided—just as we guide—here also ... realizing we are probably guided ... as well.

Leo: Are you given information about the eventual purpose of this guidance will be?

Grandpa: We are learning; it's all learning ... it's all ... the oneness, the oneness must learn. It's all consciousness; it's all ... humans do not understand.

Leo: Thank you very much. We appreciate that. Now we'll take other questions.

Matt: I have a question. Are you aware we want to take Stan up to Robert's cabin in about a month?

Grandpa: Yes.

Matt: And would you accept our invitation to come there as well?

Grandpa: We have not decided. We believe this is a good idea for Stan.

Matt: You believe it is a good idea?

Grandpa: Yes—for Stan—he must get away.

Matt: And would you let us know somehow or guide us in some way of how many cameras to bring? How many—

Grandpa: We have not decided. It is not necessary, but we are considering it.

Matt: I believe that it's necessary for others to understand that what's happening to Stan is real. Otherwise humans will not believe that it's real.

Grandpa: You believe [it's necessary] because you do not understand. You think this is the right way but do not understand. Baby steps! Baby steps ... baby steps.

Matt: I understand that you want to have "baby steps," but at the same time, in the past, say, ten years on this planet, many, many appearances from either your group or others have been photographed all over this planet. Yet many of the people here do not believe that it's real—that it's all a hoax.

Grandpa: Hmm ... it is about ... it is about consciousness. It is about ... *word* ... thorough ... to create acceptance you must be ... worked on in small steps ... small steps. If things were to happen suddenly, it would ... be detrimental. It would erase the work that has been done. Races cannot jump into things. They must experience it ... *word* ... they must experience things for themselves. It is easier to accept that way. Things cannot be rushed ... or they will not work. Do you understand?

Matt: I don't understand how it's different than what's been going on for the past ten years all over the planet, with people seeing ships and whatnot—what difference does it make if you were to—

Grandpa: (*Interrupting*) At this point it ...

Leo: ... would to you! (*Laughter*)

Grandpa: At this point, people would not believe even if we allowed it. It makes no difference. Hmm ... we have not decided. That is that.

Matt: So what would be your recommendation—should we not bother going through [unintelligible]?

Grandpa: I did not say that. We have not decided.

Matt: Will you communicate to us somehow—when you have made the decision?

Grandpa: Hmm ... we have not decided. We will let you know.

Matt: Thank you.

RM: Some of the equations you've given Stan, do some of those equations show us how to get free energy?

Grandpa: It is a ... *word* ... breadcrumb ... it is ... pointing in the right direction. We do not yet want humans to have ... technology like that. They are dangerous. There is great energy in ... some of this technology that could be used to destroy themselves. Humans are like children with fire. Until the enlightenment happens, it is too dangerous. This will not be allowed—humans will not be allowed ... out there ... until we know that they can handle themselves ... in a manner that is not ... dangerous to others ... and themselves!

RN: Is there more than one race working with you and observing Stan?

Grandpa: Yes!

RN: Can you maybe tell us how many?

Grandpa: No. It is not allowed. Some do not want to be known.

RN: Thank you.

195

V: Earlier today I had a regression with Dr. Sprinkle. From that regression came some memories of possible experiences. Were those experiences from your race?

Grandpa: Yes. Most ... not all.

V: Were there more than two?

Grandpa: Yes. There were more ... all working together.

V: On my recurring dream that I had as a child—were they dreams or were they real?

Grandpa: They were ... they were ... not us. But we try to ... collectively ... always try to make minimal impact, but there is a reason we do. But there are some that do what they do despite ... us for their own agenda but ... we are not them. Our intentions are to ... enlighten and to help ... that is why ... there is no fear ... in your experiences.

V: Have Stan and I ever been abducted at the same time?

Grandpa: Yes! You know this.

V: "Audrey" had said in a message once that they/you were baffled and amazed at how quickly Stan and I found one another.

Grandpa: It was not intended. It was not intended for you to know so soon. We were concerned that it would be hard for you to accept. Interestingly, it was hard for Stan to accept and easy for you to accept. We find that baffling.

H: Is Victoria that woman that Stan remembers from up there?

Grandpa: It is not important.

H: With all due respect, Grandpa, I hope you know how frustrating that answer is to us.

Grandpa: I say that because you know the answer.

H: No. If we knew the answer we wouldn't ask.

Grandpa: You must trust your instincts. It is a lesson.

H: Wouldn't it just be easier to give us a straight answer?

Grandpa: You know the answer. It is a lesson. Yes!

H: Thank you.

V: The answer is yes?

H: That is not the answer that we had before, though.

V: Was there more than one woman involved?

Grandpa: Yes.

V: It's getting there! (*Laughter*)

Grandpa: There were three. Two of them ... four. Three of them are in this room.

V: (*Misunderstanding*) Still learning?

Grandpa: Three of them are in this room. There were four. Three of them are in this room.

H: And who is that?

Grandpa: Who is *that*?

H: Who are the three?

Grandpa: (*Giggling*) It is not important.

Leo: (*Laughter*)

V: Is Heidi one?

Grandpa: That is correct.

V: Is Lisa one?

Grandpa: That is correct.

V: Is Victoria one?

Grandpa: (*Wearily*) That is correct.

V: There you go!

H: You had said that I had one offspring and did I want more ... I do not mind being the donor of offspring up there, but ... next time I'd like dinner and a movie.

Group: (*Laughter*)

H: And I have to credit that witty remark to Claude!

Group: (*Raucous laughter*)

Grandpa: (*Apparently confused*) ... movie ...

Leo: It's a human term having to do with sexual relations.

Grandpa: Hu- humor ... ?

Leo: Yes.

V: Human humor. Is there any way that the ET's would be willing to use a device similar to what we call a "web cam" that could be linked to our Internet for our Earth people to view video signals, perhaps of the interior or exterior of your craft, as well as your home cities, so that we could enable people in the comfort of their own homes to get used to seeing what your planet is like, what your craft are like, and what you are like?

Grandpa: Hmm ... it is not necessary—*maybe* in the future. That is not necessary; humans are not ready ... it will be sometime, but in the future. There will be no need. When the shift happens … none of this will be necessary … this proof, this …

H: Grandpa, we know that communication with Audrey drops off in the fall and winter months [unintelligible]. When we try to communicate with your people telepathically, do you receive those messages or do you have to be monitoring us to receive them or, how does—

Grandpa: Hmm … I just think that it's hard. Communication is … Humans do not communicate the way … humans … are not capable. In order to … communicate, we must be close. Between ourselves, it does not matter. Everything is … in the same space. Humans can't understand time, space—everything is in the singularity but … it is hard to explain. You would not understand. With humans, we must be close. It is difficult so that is why … we communicate the way we do. You call it … *word* …

V: Telepathic.

Grandpa: Yes, but—

H: Audrey?

Grandpa: Audrey.

H: So the only way that we can communicate with you is to wait for communication from Audrey?

Grandpa: Hmm … sometimes. Stan is different. We communicate through Stan. Stan is designed … we use Stan among other things, but … if we are close, there is no need to communicate—we know.

H: And why is it that during the fall and winter months, it's difficult for you to come through?

Grandpa: We have discussed this. It is matter of alignment.

RN: Do you live on your planet of origin?

Grandpa: We live many places. We have … [unintelligible] like most very advanced races, we have [unintelligible]. We are part of the collective, collective conscious. We are … part of the council. We are … yes, we live there and other places.

RN: Are there human members of your council?

Grandpa: Hmm … no. There are … human members—one or two— lower … not … they have been chosen, but … we do not know them.

RN: Those at St. Cajetan's in January at Stan's presentation, were they your race?

Grandpa: (*Sly smile*) It is not important.

V: Recently in reading the transcripts from previous regressions of Stan's, a woman had driven up to Stan and told him that they should "reevaluate"—in order to understand—they should reevaluate the Rosen Bridge. Are you familiar with that communication?

Grandpa: Yes.

V: Who is the we—or they? Is it science in general or those—

Grandpa: That was not us. Not us. That is ... Stan is ... inside, inside, inside ... inside help. That is ... government. Good ... good government. That is ... in general ... we believe, in general. Everybody needs to reevaluate ... they know.

CS: So first-rate science needs to reevaluate that idea?

Grandpa: Yes! Sure. But everybody needs to reevaluate that, to understand it. That is just a small part. It is ... you do not understand; it is more than that.

H: Grandpa, there was a woman that Stan met as a child ...

Grandpa: Yes.

H: ... whom you told us was his offspring.

Grandpa: Yes!

H: Has he seen that woman since puberty? Has ... she ... been in his life—

Grandpa: Yes! He has not seen her, but she has seen him. She monitors Stan.

H: Do all of his offspring have the ability to—

Grandpa: No. She is different. She plays an active ... (*slowly*) active ... active ... active role. She is of the enlightenment. She is ... very intelligent!

H: So she is a hybrid that's able to be here? on Earth?

Grandpa: Hmm ... hybrid ... yes—hybrid ... yes. She monitors ... each of you; Lisa, you, Victoria, everybody ... involved. You have met her ... briefly.

H: I have?

Grandpa: Yes ... recently.

V: (*Gasping*) This last weekend at the Sand Dunes?

Grandpa: It is not important.

V: Do you know who broke into Stan and Lisa's house ... when the UFO file was taken and—

Grandpa: We are aware.

V: Do you know who broke in?

Grandpa: We believe ... we believe is they that do not want this out ... same as ... we have said that Stan needs ... not to walk on his own. There are those that are scared ... because they know their ... their reign is coming to an end.

V: Are they human?

Grandpa: Yes!

Matt: There is a certain amount of danger involved with our bringing this forward.

Grandpa: Once it's forward, there will be no danger. There is danger in not bringing this forward.

Matt: I understand that. But there are also those who would silence us from their own fear and stop us from bringing this information out.

Grandpa: They will try that to stop you from bringing this information out ... but once the information is out, they cannot stop you.

Matt: I understand that, but in the meantime, is there some way that you can protect those of us who are—

Grandpa: We are—we are [unintelligible]. There are those inside that are helping. We [unintelligible] ... *word* ... there have been many times where your life has been in danger but it has been stopped. Stan has ... *word* ... he has ... it seems he has guidance above us.

V: Interventional guidance?

Grandpa: Hmm ... we don't know, but it seems there is more to this than—

V: Is there punishment for those who [unintelligible]?

Grandpa: They are just lost.

V: When Stan was recently shot at, was it more to scare him than actually to hurt him?

Grandpa: No. Not to scare him. We intervened.

V: With the orb in the tree? With the light in the tree?

Grandpa: Among other things.

Matt: So we're not to be afraid to go forward with this—

Grandpa: No, you must. For safety reasons you must ... go forward!

RS: I have a question. You mentioned earlier that there were several races here and are still here [unintelligible] on this planet. Now, are those races physical races?

Grandpa: Yes!

RS: All of them.

Grandpa: Some are ... some are not. Many are—humans do not understand. There are multiple, (slowly) *multiple* ... many *multiple* layers ... layers ... in this [unintelligible].

RS: OK.

Grandpa: But ... you are three-dimensional. Physical reality has ... others living. Others that are ... the government ... government knows and acknowledges, and there are others that do not ...but ... there are others that do not know.

RS: Thank you.

SD: I'm trying to teach a theory of universal law to people upon—

Grandpa: You do not know universal law. How can you teach a theory on universal law?

SD: Can you give us—

Grandpa: When it is time. You will learn ... when the shift happens, it will be ... necessary. You will be part of it all.

Matt: We need to understand more about this shift. Is this a gradual change, or is it a sudden—

Grandpa: We do not know. It can be both. It can be gradual or sudden. So far it is gradual. But we believe it will reach a point where it will ... like ... falling off a cliff—

Matt: Sudden? Natural?

Grandpa: Hmm ... walk ... up ... hill ...

V: Incline?

Grandpa: Incline, yes. Gradually, suddenly, it just happens. We believe that is how this will happen, and it is not just individual—not individual.

RS: Collectively?

Grandpa: Yes.

Matt: Is there then no real significance to December twenty-first, 2012 ... in our time?

Grandpa: Hmm … it is not allowed. There will be things that take place and … some … are not good … Stan … talks about … it cannot be fixed; it must be replaced. Do you understand?

Matt: Not exactly.

Grandpa: There will be things that transpire … within the next few years that … are going to be uncomfortable and … meant to be devastating. But … as a race … the race will have two choices. One to push through, or one not to and perish. And we hope that your race will push through.

Matt: Then why are these changes happening now?

Grandpa: Hmm … humans have … been guided incorrectly, and … it is just part of the shift. It has to happen, though. It has to be … *word* …

Matt: Chaos?

Grandpa: Chaos before … there is … *word* … it is like shaking a carpet … to get the dust out—Stan's … Stan's thoughts.

H: Should we be collecting food or supplies or—

Grandpa: There is nothing you can do … you'll know … you'll know … instinctively … what to do when the time comes. You will know. You will band together. When it is over, the human race will go on again. Unfortunately, we cannot get involved; we can only guide. We are … we are sure that … things will go the right way, but … it appears to us that … there are those … that are trying … the … human race is being … there are few that are controlling the many, and this must change.

CS: This is Claude and I have a question.

Grandpa: Yes, Claude.

CS: I had trouble finishing my second question—

Grandpa: Yes.

CS: And I wonder—is that because there are things that should not come out—

Grandpa: No, you are stuck because … you are having a hard time focusing. It will happen. You must wait, though. It is a … *word* … it is … *timing*. There are things in the book that are important but … timing …

CS: Not time yet?

Grandpa: It's not the time yet, no.

Matt: Are we on the right track with ABC Network? Should we be involved with a major radio network like this, or should we be independent of people like that?

Grandpa: What you are doing is fine. Little pieces here and there can only help ... advertising ... (slowly) *advertise* ... advertise. You must be careful, though. You must not give full control. Just what you are doing; you are on the right track. It is our experience that ... what we have learned that ... it is not about enlightenment with ... these groups. It is ... these groups are ... they have their agenda, some of which are ... controlled by ... Paul ... by ...

PW: Cabal?

Grandpa: Yeah, cabal. Cabal ... but ... they are losing ground also. There is a war! There is a war between these ... factions ... (slowly) *factions*, cabals ... there are ... those that want this out, and there are those that don't. There are those that are naturally being enlightened because they have no choice. And ... this is wonderful. We are excited. But there are those that will do anything to stop it, and they will stop at nothing. We are closely monitoring these—we believe it will happen despite them. There will be [those] that will not be enlightened, and they will perish. But ... it is our hope that you are strong.

V: Will Stan's book be a bestseller, and will it help to—will people believe what they are reading as true experiences?

Grandpa: Stan's book will help. There is more than just Stan's book. We do not know. We do not—well ... we cannot get involved. But we believe it will do well.

H: Are you familiar with the Extraterrestrial Commission?

Grandpa: Yes!

H: In Denver?

Grandpa: Yes!

H: What is your feeling on the *Boo* video being aired.

Grandpa: It is fine. It is ... part of the process. It is ... to spark—Stan's words—spark ... spark ... it is to spark ...

H: Interest?

Grandpa: Interest—yes—thank you. Interest ... and ... it is part of it. There will be those that do not believe, no matter what. And there will be those that do. But ... everyone here must be careful. Easily, this can be looked at ... as if it were religion—religion—religion. It is not meant to be that. It is ... this is enlightenment. This is not ...

RS: In place of their God?

Grandpa: Exactly. Exactly. Thank you.

V: Is it a good idea to reapply for the ballot initiative to allow more time for signatures?

Grandpa: Yes! This is ... part of—you must understand—and ... Jeff? Jeff has also been brought into this, not by accident. This is all part of the process. This is to ... bring attention to... this. Attention to all of *this*. Stan ... others ... will come together—this is part of the process. This was designed to ... expand outward. This is all part of that process. Stan ... everybody ... involved in this here ... is an important part.

V: May I ask what my part is?

Grandpa: You are to support Stan. You are to guide Stan and Lisa. You are ... smart in business. You are ... more. You have been involved with Stan for a very long time. You do not understand. Everyone here ... is ... part of ... the process. You must support each other and, in turn, be supported. It will happen. This is a good thing.

RN: It sounds like you have more confidence now than in previous years that we will have taken the appropriate steps and actions to allow us to survive beyond 2012.

Grandpa: Yes. You should be proud of yourselves. You have done well. Stan has done well. Stay steadfast ... (slowly) *stead* ... stead*fast*. Do what you are doing. Do not be afraid to stand your ground ... Claude will be ... Claude will be hit ... hit hard ... but he must stay strong. Stan will be hit hard, but he must stay strong. Opinions will change. Opinions are already changing. The people who are ... trying to discredit Stan are being looked at and laughed at, and ... that is part of the process. As more people accept, more things will ... come forward. And it will be a snowball effect. And that's when the shift can happen.

LR: OK. I have a question for you. I have a memory of a being handing me three babies, two of which I could hold, one I could not. Is it a dream (*tearfully*) or a real memory?

Grandpa: You have ... soon it will be time. You have been involved also. And ... you must go with your instincts. We have ... we have kept things ... from you ... to ... minimalize ... (slowly) *minimalize...* minimal ...

Group: Minimize?

Grandpa: Minimize stress, pain. You are important. You have fears unfounded. There is nothing to worry about. Nothing ... nothing to worry about. It is all part of the process. Everything that is happening is part of the process. It is all part of the greater good. Even what we are doing ... we realize now it's probably guided by something higher—nonphysical. We are studying this. It is all part of the—

H: Often, Grandpa, it is much more stressful for humans to live with what is *not* known than to know the truth.

Grandpa: It is ... now sometimes, it is ... harder for ... humans to know the truth.

H: And sometimes it is harder for us *not* to.

Grandpa: Hmm ... you don't understand.

RN: But there are many of us who do want to understand, who understand more now.

Grandpa: And you will. You will understand. Baby steps! You are not ready. Despite what you say, you are not ready. You know this. It is ... *word* ... like ... like ... Stan ... from Stan, *word* ... like ... letting a child drive a car ... without knowing what he is doing. Do you understand?

RN: Not completely. We understand more than many. But we want to understand more.

Grandpa: It is a process! You must learn a little bit at a time to ... there is ... there is no way for you to understand ... unless it is in small steps. The things you see around you are ... such a small part of what's really there. You would be overwhelmed. And ... in order to stop this from happening, there must be baby steps.

V: Lisa awoke several years ago with a triangle on the back of her head. What is that?

Grandpa: Monitor ... monitor ...

V: Does she have an implant that was placed there?

Grandpa: Yes … not there—not anymore.

V: Do I have an implant?

Grandpa: Yes. Do not be afraid. This is … to monitor. This is to … not … not … aggressive. Passive.

H: Do I have one …

Grandpa: Yes!

H: … in my ear? Is it in my ear—is that a memory? Is that memory accurate?

Grandpa: Hmm … one … there. Not anymore. Here … (*pointing*) back … here.

H: Do I have one elsewhere now?

Grandpa: Yes.

H: Is this just as a protection thing?

Grandpa: Monitor … monitor, passive. We must monitor. You guys … guys … guys … guys … guys—Stan's words—(slowly) *you guys* … strange … memory … strange. Everyone involved … must be monitored. Everyone involved is important.

RS: Does everyone in this room have a device?

Grandpa: Um … everyone … except you.

RS: Hmm. Thank you.

Group: (*Laughter*)

H: Good! Something to look forward to!

Group: (*Laughter*)

RM: Grandpa, did you just say that I have a device in me?

Grandpa: Yes … not fear, no fear … just … monitor. Monitor. Richard does not have one because he is not … he is involved as a friend. He is … will be involved later. (*To Robert*) Yes—you have … just monitor, that's it. Stan more … he is different. Stan is special.

CS: The next time Stan has a device moving around, interacting like the one recently, would it be helpful for us to try to measure it—

Grandpa: No.

CS: No?

Grandpa: Not allowed.

SD: When was the last time you took Stan from his house?

Grandpa: Recently.

SD: OK. Can you talk about that at all, or—

Grandpa: Hmm … no. UFO … a lot of you call "UFO."

SD: Have I been taken by any groups?

Grandpa: Yes. Not recently.

Matt: Have I been taken by a group?

Grandpa: Yes.

Matt: When was this?

Grandpa: You remember. Think.

Matt: I have no memory of it.

Grandpa: Briefly. Nothing ... nothing major.

Matt: Recently?

Grandpa: No.

PW: Grandpa, you mentioned at a previous meeting that I was taken as well, as Heidi and Lisa have.

Grandpa: We do not ... we do not ... leave memory. Minimal impact. There are some that ... we cannot help. There are some that are ... higher ... more advanced ... we cannot control if you remember or not.

PW: Did your people take me, Grandpa, or was it someone else?

Grandpa: Hmm ... someone else ... but still ... within ... council— what you call ... council.

H: Grandpa, I know that you will take our memory away of these experiences so that we won't be traumatized.

Grandpa: Yes.

H: When it is after the fact, at least for me personally, I don't believe it would be at all traumatic to remember, and I very much desire to remember.

Grandpa: Hmm ... yeah, but we will not. Not unless it is necessary. There are some things that might be painful but ... this is not meant to be ... with us. This is not meant to be a traumatic ... (slowly) *traumatic* ... traumatizing ... traumatic experience. This is about the enlightenment. This is—

H: And I believe that.

Grandpa: We want ... as you help us; we want ... to return the ... Stan's word—favor ... (slowly) *favor*. There is more at stake here than you realize. Do not worry about ... the memories. Do not wor— there is a bigger picture ... picture, picture at stake. Do you understand?

H: Yes. It does help us. It would help us when we talk to other people and say this is real because we have never even—

Grandpa: You can ... if need be you can recall these memories. You are starting to ... you are also starting to also recover part—a little bit as ... a little bit at a time [unintelligible].

LR: We recently had a paranormal investigator come to our house.

Grandpa: Yes.

LR: And without naming names on the video and everything, were you there?

Grandpa: Yes.

LR: And did you communicate?

Grandpa: Not ... personally, but ... yes. There were those that were there, yes.

LR: Because they were very familiar to most of us there.

Grandpa: Yes.

LR: It was like talking to you.

Grandpa: Yes, it was ... me—there are many me's. Many, understand ... many ... many ...

LR: Many of your kind?

Grandpa: There are three ... . um ... through Stan—three.

LR: You, Audrey, and one other?

Grandpa: No. Three ... council that ... through Stan.

RS: Communicate through Stan?

Grandpa: Yes.

V: So you aren't the only one we talk to through Stan?

Grandpa: No. There are three ... allowed.

H: And we call all three of you "Grandpa"?

Grandpa: Hmm ... that is fine.

RN: Are there more than the three of you when you communicate?

Grandpa: Yes!

RN: Is it this communication ...

Grandpa: Yes, you got it.

RN: ... a multiple ... thing?

Grandpa: M-hmm ... sometimes ... sometimes there are more; sometimes there are less; it depends. Now there are less.

V: Is it those same three who communicate as "Audrey" ... when it is your group?

Grandpa: It depends.

V: On ... ?

Grandpa: The time. If it is important there are more; if it is not important there are less. It is a matter of ... alignment. It is hard to communicate that way, but it is the only way. You will understand ... we are not necessarily ... that voice, but ... it is familiar with Stan ... to Stan.

V: Can you communicate telepathically to us?

Grandpa: To Stan.

V: When the shift occurs—

Grandpa: Yes.

SD: Are you aware of a hole in the Arctic—a hole in the Earth?

Grandpa: Earth ... Earth ... Earth ... ozone?

SD: No, not a hole in the ozone, a physical hole.

Grandpa: Lizards ... um ... Paul.

Group: (Laughter)

RS: All of it true!

Grandpa: There are many holes; there are many things. There is just—

SD: But there's not like somewhere where there's a hole or something—

Grandpa: There are civilizations ... that are ... a few places. It is not allowed.

V: How many on your council? What is the number?

Grandpa: It varies.

V: And are they all from your planet?

Grandpa: No.

V: And none are from Earth?

Grandpa: No. Some are on Earth, but they are not from Earth.

RS: Grandpa, has everyone in this room been taken ... or visited?

Grandpa: Visited, yes. Not taken. Most have, but ... just visited—some more than others. There are some that have been taken several times.

H: Does the "New World Order" defy Universal Law?

Grandpa: Yes.

H: Is their punishment part of the law?

Grandpa: They're just lost. They'll punish themselves. Humans do not understand—they do not need punish—they'll punish themselves ... by what they do. They are missing the point. They will be lost because of their own doing. That is punishment enough.

SD: Do you look at humans as a two-in-one—do we have a soul and have a biological—

Grandpa: Humans are ... humans are different. Humans are ... more than they know. We are studying this.

SD: Are we two entities in one, though—biological and a spiritual?

Grandpa: You would not understand. Everything is ... connected. Everything has both, but ... you do not understand—you would not understand.

RN: Are you familiar with Dr. Emoto's study of water and the change of molecular structure in the water?

Grandpa: Somewhat.

RN: The reason I ask is [because] you continually call us "water vessels" and much more.

Grandpa: Yes.

RN: Do we have the ability to, through writing—on us—change our molecular structure?

Grandpa: Hmm ... too complicated. You would not understand. Humans ... global consciousness as a whole ... whole—Stan's word—global consciousness as a whole can manipulate ... the reality. They have the ability, part of it deity ... part of the oneness. Humans do not know who they are. Stan is getting fatigued; we must end. Two more questions.

Leo: (To group): Two more questions?

SD: In space, there is dark energy, and there's light energy ...

Grandpa: In space there is more than dark and light energy.

SD: OK. So there is more.

Grandpa: There is more. Humans ... there is no "space" in space.

SD: In the manifesting of those energies, or the creation of those energies, does it have anything to do with the life—the consciousness of the beings that are within—

Grandpa: There are many consciousnesses—all part of the oneness.

RM: Are you part of the white light?

Grandpa: We are part of the oneness. There are many types of light.

V: We were all at one at one time and put off from the oneness?

Grandpa: Hmm ... that is what we understand, yes. We are studying that. One more question.

V: Is that considered the "atonement," the "At One"?

Grandpa: Hmm ... too complicated. Humans do not understand.

V: Thank you.

RN: Will we see you tonight?

Grandpa: It is not necessary.

H: (*Pleading*) Come on?

Group: (*Laughter*)

RS: Put on a little light show!

Grandpa: Stan is tired.

Leo: Stan is tired, and we thank you, Grandpa, for being with us.

Group: Thank you!

Leo: Thank you very much. We'd like to talk with you later. And as Grandpa departs, when you're ready, Stan, you can return back to the normal state ... feeling fine, feeling good. Let yourself return to the normal state, feeling fine, feeling good.

End of regression session five

Many of the revelations provided by Grandpa the Orion hinted that humanity was far greater than it realized. Never was he able to reveal why. In parallel with that notion, he repeatedly warned that we were at a crossroads—that our existence on this planet could go one way or the other. Shortly after this last session, a series of events occurred, spoken of in *Answers*. In the chapters addressing the "Timeline Wars," I go into detail about why Grandpa felt humanity could be destroyed and also about what humanity did in late September and early October of 2008 to begin what Grandpa called the "Shift."

What you will also discover is information that even the Orions weren't aware of but tried to research regarding the wonder of humanity. In a few references, Grandpa related that my soul was sent here to play a part in what is now unfolding—and that part of the unfolding included the possibility of cataclysmic human failure. Indeed, in *Answers*, I go into further detail as to what the Orions saw in our future and what the

Possum Lady showed me that left me sobbing at night, crumpled on the floor with grief. But now I can speak of a great joy, an unspeakable hope for all of humanity. That hope is addressed in *Answers*.

Grandpa also spoke of a time that might occur when he would allow us to see him. It is only now that I realize why he waited. There have been appearances, but not by Grandpa. Something even greater than Grandpa. You will see pictures in *Answers*, and you will get answers as to why Grandpa used such restraint in dealing with our inner circle. Humans are about to realize the miracle that we are in the cosmos. I am both proud and rejoiceful to have been born on Earth. And when you are done reading both of these books, I believe you will share this with me. Something wonderful is happening, and it's high time you are allowed to see it. It will require one small effort on your part. You will have to open your eyes. "Those that have eyes to see, let them see."

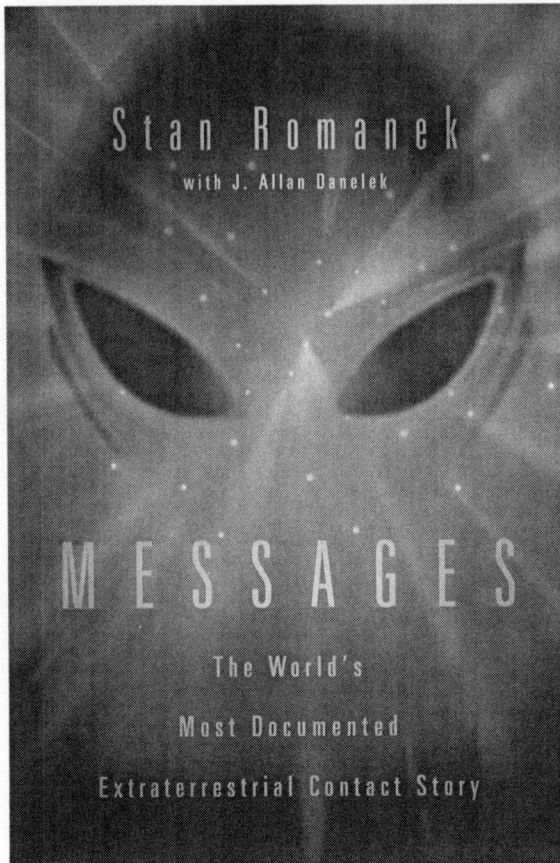

Stan Romanek
with J. Allan Danelek

MESSAGES

The World's

Most Documented

Extraterrestrial Contact Story

MESSAGES:
The World's Most Documented Extraterrestrial Contact Story

An exciting true story, but far more than that, *Messages* will answer a lot of questions about whether or not we are alone, and why "they" are here. It tells the first person story of a contactee.

Order at www.StanRomanek.com

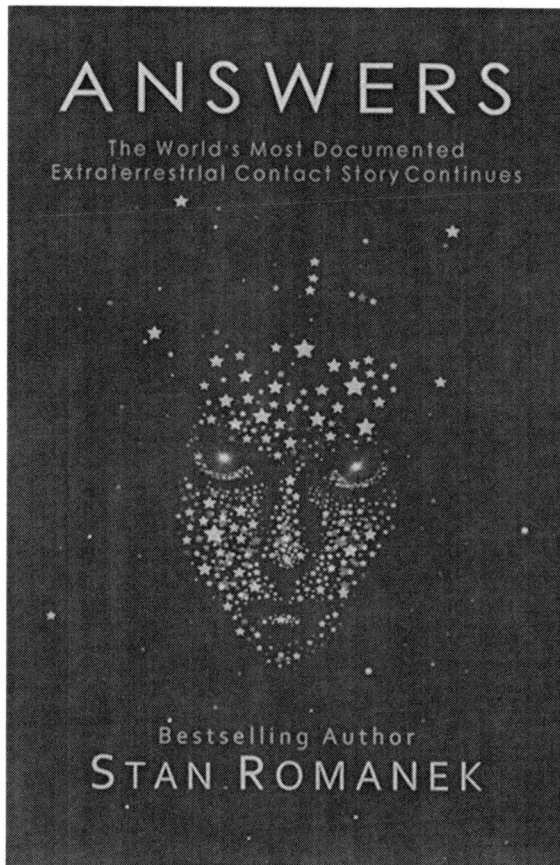

ANSWERS
The World's Most Documented Extraterrestrial Contact Story Continues

Bestselling Author
STAN ROMANEK

ANSWERS:
The World's Most Documented Extraterrestrial Contact Story Continues

More than the sequel to the best-seller *Messages*, *Answers* also addresses the worldwide concerns around 2012 and planetary calamity, drawing from actual experiences of Stan Romanek with ETs giving him glimpses of the future of planet Earth, comparing those with corroborating information from governmental whistleblowers and their use of future-viewing technologies.

Order at www.StanRomanek.com

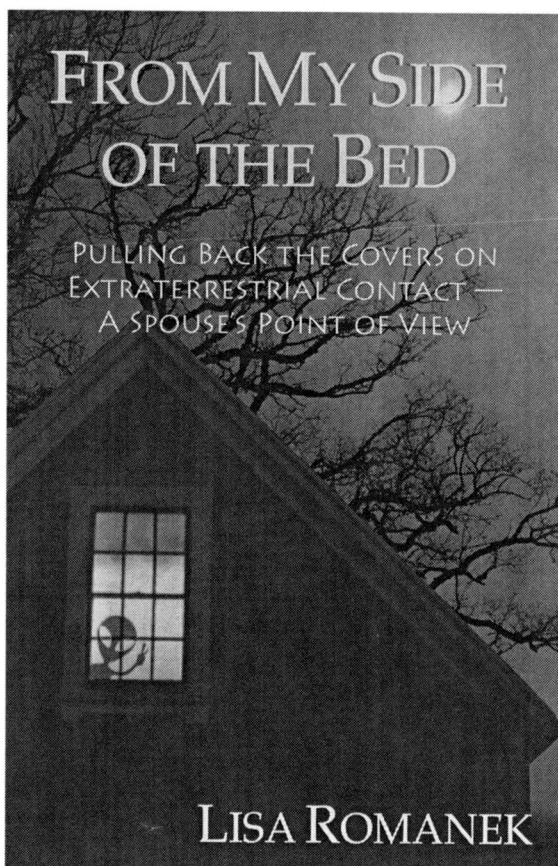

FROM MY SIDE OF THE BED
PULLING BACK THE COVERS ON EXTRATERRESTRIAL CONTACT—
A SPOUSE'S POINT OF VIEW

LISA ROMANEK

FROM MY SIDE OF THE BED:
Pulling Back the Covers on Extraterrestrial Contact—
A Spouse's Point of View

A true account of extraterrestrial contact—but for the first time from a spouse's point of view, and with a humorous edge. More than a "how-to" book in coping, Lisa addresses deep and biting questions that have rarely been addressed in public dialog. Families are being torn apart by the trauma, shame, and stress of discovering their lives have been invaded by uninvited "guests" who seem to be ignorant of the chaos they leave behind.

Order at www.romaneks.com

About the Author

Stan Romanek's extraterrestrial visitations began in 1968 as a five-year-old child. However those memories stayed buried until September of 2001, flooding his consciousness after the first of his numerous alien abductions. In his bestselling book, *Messages: The World's Most Documented Extraterrestrial Contact Story*, those abductions are documented like no other case in history. Romanek's case has stirred controversy in the scientific community because of the highly advanced mathematical equations he has brought forth during sleep and under hypnosis, pointing to wormhole technology and travel across time and space.

Stan travels the country, presenting lectures and sharing his story of extraterrestrial visitations,where he evokes his emerging understanding of the Oneness, our connectivity to all life in the Universe, and humanity's lost knowledge of who we really are.

Stan has been a frequent guest on hundreds of radio and TV programs such as *Larry King Live*, *Fox and Friends*, and ABC's *Primetime* program, "The Outsiders."

The multitude of unexplainable events involving Stan Romanek and hundreds of witnesses continue on a regular basis. The sheer volume of documents, video, audio, and photographs is unparalleled. Stan's ongoing story continues to become more complex with each twist and turn. The overriding conclusion from this continuing saga is that there is much for humanity to learn.

Stan lives in Colorado with his wife Lisa where together they work diligently to fulfill the duty given to Stan by the Orions: that of a messenger.

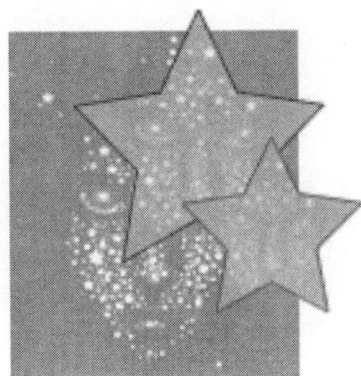

9 781893 641068